Animal Behavior

Second Edition

Animal Behavior

SECOND EDITION

An Introduction to Behavioral Mechanisms, Development, and Ecology

MARK RIDLEY

Departments of Anthropology and Biology
Emory University, Atlanta, Georgia

Boston

Blackwell Scientific Publications

Oxford • London • Edinburgh
Melbourne • Paris • Berlin • Vienna

Blackwell Scientific Publications

EDITORIAL OFFICES:

238 Main Street, Cambridge, Massachusetts 02142, USA

Osney Mead, Oxford OX2 0EL, England

25 John Street, London WC1N 2BL, England

23 Ainslie Place, Edinburgh EH3 6AJ, Scotland

54 University Street, Carlton, Victoria 3053, Australia

Arnette SA, 1 rue de Lille, 75007 Paris, France

Blackwell-Wissenschaft, Kurfürstendamm 57, 10707 Berlin, Germany

Blackwell MZV, Feldgasse 13, A-1238 Vienna, Austria

DISTRIBUTORS:

USA
Blackwell Scientific Publications
238 Main Street
Cambridge, Massachusetts 02142
(Telephone orders: 800-759-6102 or
617-876-7000)

Canada
Oxford University Press
70 Wynford Drive
Don Mills, Ontario M3C 1J9
(Telephone orders: 416-441-2941)

Australia
Blackwell Scientific Publications (Australia)
Pty Ltd
54 University Street
Carlton, Victoria 3053
(Telephone orders: 03-347-5552)

Outside North America and Australia
Blackwell Scientific Publications, Ltd.
c/o Marston Book Services, Ltd.
P.O. Box 87
Oxford OX2 ODT, England
(Telephone orders: 44-865-791155)

Acquisitions: Jane Humphreys
Development: Gail S. Segal
Production: Kathleen Grimes
Design: Joyce C. Weston
Typeset by Huron Valley Graphics
Printed and bound by BookCrafters,
Chelsea, MI

© 1995 by Blackwell Scientific Publications

Printed in the United States of America

95 96 97 98 5 4 3 2 1

Library of Congress Cataloging in Publication Data

Ridley, Mark.
    Animal Behavior: 2nd ed./ Mark Ridley
p. cm.
Includes bibliographical references and index
ISBN: 0-86542-390-3
1. Animal behavior
for Library of Congress   94-21727
                                            CIP

Contents

CONTENTS

Preface

Animal behavior may be the least fragmented subject in modern biology, and modern psychobiology. The people who study animal behavior like to understand, and make use of, all the main research strategies. They think of animals as machines with fascinating mechanics, as organisms with complex life histories, and as the end-products of Darwinian natural selection, that have evolved from ancestors who themselves lived and behaved. The full science spans from microscopic investigations of genetic and neurophysiological mechanisms, which are done using artificial systems in the lab, through "non-invasive" experimental studies of the behavior of whole organisms, to work that is more like natural history and is done outside, in the forests, fields, and oceans. One of the pleasures of updating this book for a second edition has been for me to follow the advances in research areas as apparently far apart as the brain mechanisms of perception—we are living in the golden age of neurophysiology—or bird song, or the hormonal mediation of stress, or the abstract conditions needed for signals to be honest: all these topics turn out to be unified in the broad science of animal behavior that Tinbergen (and others) have inspired.

The book is an essay in concise introduction. I have aimed to introduce the main principles of animal behavior, and wherever possible I have done so using examples—as most of us find it easier to pick up ideas for the first time in the form of a concrete example rather than an abstract discussion. I have not ignored unsettled questions (indeed I have tried to draw attention to them, to indicate where the science is heading), but I have also emphasized positive knowledge, in the belief that an introductory work should say something about what is known in addition to how we should think about behavior scientifically and what research questions are currently being researched. When a classic example can be used to explain a principle, I have not hesitated to reach for it: the book is, after all, not an encyclopedia or a refresher course, but an introduction. I hope it will lead some readers further into the subject, and I have used the "further reading" sections at the end of each chapter to signpost the way on.

CHAPTER 1

Questions About Animal

Behavior

What kinds of questions can be scientifically asked, and answered, about animal behavior? We begin with this question and move on to see how objective 'units' of behavior can be recognized, and how an apparently complex feat of behavior such as the spider's web can be broken into an intelligible sequence of simpler behavioral units.

1.1   Why do animals behave?

All living species of ants live in "social" groups, a group being a colony that contains a number of individual ants. Ants characteristically cooperate with other members of their own colony. The may cooperate in catching and transporting their prey or other resources to the nest, they may cooperatively look after the eggs and larval ants in the nest, and they may cooperatively defend the colony from its enemies. Ant colonies, like all living things, have enemies: predators that would feed on the colony; parasites that would exploit it; and other enemies, particularly other colonies of ants, that would take over or steal the resources of the nest. If the colony is to survive, the enemies must be deterred. Most ant colonies possess a special caste of soldiers to do just that. The soldiers of many species have enlarged mouthparts for crushing or snipping, together with a large head to house the muscles that power the mouth-parts. Ants use chemical defenses too: they may spray poisons or glues. In the case of the Malaysian seed-gathering ("harvester") ant (closely related to the species *Camponotus saundersi,* which has been studied by Ulrich and Eleanore Maschwitz), glues are not merely sprayed; when an ant is disturbed it actively explodes, covering any nearby enemies with a fallout of glue. This species has an enormously enlarged mandibular gland, where the glue is manufactured (Figure 1.1), many times the size of that found in other ant species. A harvester ant in trouble, for instance when it is fighting another ant, explodes itself by contracting the muscles of its abdomen with sufficient strength to split its cuticle. The sticky contents of the mandibular gland then burst out (Figure 1.2) and the glue entangles and immobilizes the victim. The self-sacrifice of the exploded ant should benefit its colony by knocking enemies out of the action, but at the cost of its own life.

Why should ants sacrifice themselves in this way? Moreover, this is

Figure 1.1

The harvester ant *Camponotus* (species near *saundersi*) possesses enormously enlarged mandibular glands. In a more typical ant species, such as the *Iridomyrmex humilis* illustrated in Figure 7.4 (p. 172), the mandibular gland occupies a relatively small part of the head; in this species of *Camponotus* the paired gland extends into the abdomen, where it takes up much of the space. *(After Maschwitz and Maschwitz)*

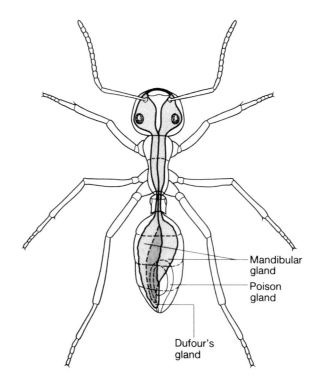

Mandibular gland
Poison gland
Dufour's gland

Figure 1.2   When a worker of this *Camponotus* species is disturbed, it contracts its abdomen until it splits; the sticky contents of its mandibular gland then explode out. The ant is held here in a pair of forceps. *(Photo: Ulrich Maschwitz)*

not the only kind of self-sacrifice to be found among ants. There is also, for instance, reproductive self-sacrifice. Typically only one individual in an ant colony—the queen—breeds; all the other colony members are sterile "worker" ants, whose business in life is to enable the queen to reproduce as much as possible. The queen is an almost static egg machine. She does not feed herself; food is brought by her workers. Nor does she defend herself from enemies; this is also done by her workers.

Each egg, after it has been laid by the queen, is carried away by the worker ants, and the eggs and young larvae are protected, fed, and reared by the workers. Within the living world as a whole, this arrangement, of many sterile individuals working to increase the reproduction of another member of the species, is exceptional. Species of ants, bees, wasps, and termites have sterile workers, as do a couple of species of aphids. Other examples have been suggested, and no doubt the list will be extended. However, in the great majority of species, no individual is sterile except by accident; all individuals are capable of reproducing (which is not to say that all individuals of the species *do* reproduce— obviously many die before maturity). Why should ants have sterile castes?

The exact opposite of self-sacrifice for others is selfish exploitation. Black-headed gulls (*Larus ridibundus*) breed in large and dense groups, also called colonies. A gull colony, unlike an ant colony, is made up of many familial reproductive units of one male, one female, and their eggs or young. The individuals of a gull colony do not cooperate—quite the reverse. They pursue their own advantage without taking any account of the advantage of other members of the colony. They will, for example, eat each others' offspring (Figure 1.3). When a gull chick hatches from its egg it is, in the words of Richard Dawkins, "small and defenceless and easy to swallow." If the chicks become separated from their parents, other gulls in the colony may well take advantage of the easy meal, and eat them. Why are some animals as apparently altruistic as the auto-destructing harvester ant, but others as ruthlessly selfish as the cannibalistic black-headed gull?

If, in a species like the black-headed gull, a young bird risks death when it goes near adults of its own species (other than its parents), it is clearly important that it should stay close to its own parents. In order to do so, it must be able to distinguish its parents from other objects in its environment, and coordinate its powers of movement with its abilities of recognition in such a way that it does not wander toward moving objects other than its parents. The parent must also be able to distinguish its chicks from other chicks, in order to decide correctly which chicks to eat and which to look after. Parents and chicks do have these abilities. The

Figure 1.3
Colonially nesting gulls are notorious cannibals. Small undefended chicks are taken in large numbers by the adult gulls of the colony. Different adult gulls specialize in hunting different kinds of food, and some individuals specialize in cannibalism.

abilities have been studied in some species of gulls and also, among other species, in mallard ducks. The mallard duckling faces a similar problem to a gull chick: its world is made up of one individual that cares for it, and many other living and nonliving objects that are at best indifferent and at worst actively, lethally, hostile. The duckling must distinguish the former from the latter, and stay near the former. Ducklings do in fact stay near their mothers, but how do they manage to do so? If we watch the mallards of a pond in the breeding season we see particular ducklings following particular adults. What process causes the duckling to grow up to follow one individual rather than another? What are the distinctions it is making, and how does it make them?

Consider another example of animal behavior. A salmon (such as the Atlantic salmon *Salmo salar,* which breeds on both sides of the North Atlantic) typically is born (hatches from an egg) in a river tributary. There is no question of parental recognition in salmon because the parents are already dead by the time the egg hatches, but the young do have another problem of recognition. After spending a few months in the river, the salmon migrate to the sea and actively swim around for a year or two, during which time they will swim in many unpredictable directions through thousands of miles. They then finally return to freshwater to breed. Their famous skill—apart from their (in human terms) heroic journey upstream to breed and then inevitably die—is to be able to return to the exact tributary in which they were born. This fact was first established quantitatively by observations of the Atlantic salmon that breed in the River Tay, in Scotland. Starting in 1903, 550 young salmon smolt in their natal river were marked with silver wire before swimming downstream. The number of marked adult salmon later returning to the Tay and to other nearby rivers was recorded. All of the 110 marked salmon that returned to any of the rivers returned to the Tay: no marked salmon had strayed to other rivers. Since then larger studies have been made. The largest experiment was on the Atlantic salmon of the River Miramuchi in New Brunswick. Stasko and his colleagues marked 174,509 smolt by fin-clipping; the 2425 adults later recovered were all taken in the natal stream. As in the River Tay experiment, 100% of the salmon homed perfectly. In other experiments, not all the adults have been recaught in their natal stream; but the stray rates are low, about 2–3% of the total number of adults. The conclusion is that salmon find their way home with high accuracy. A salmon swimming off the Atlantic coast of North America will have a choice of hundreds of freshwater outlets into the sea; even within its natal river it will have to choose correctly among many branches and subbranches in order to relocate its natal tributary. This is an astonishing ability, perhaps because a human would be incapable of it (even if we could swim underwater continually for a year or two). How do salmon find their way home?

Such questions—as why ants sacrifice themselves for their colony, whereas blackheaded gulls would sacrifice their colonies for themselves, or why a mallard duckling follows one particular adult mallard rather than another, or how a salmon finds its way home—are all examples of the questions we shall be seeking to answer in this book. They are also the kinds of questions that make up the science of animal behavior. Ethologists, behavioral ecologists, comparative psychologists, and sociobiologists (there is no universally accepted term for the whole subject) aim to discover how animals do behave, and then to ask, and answer, questions about why the animals behave the way they do.

The general question of animal behavior—"Why do animals behave?"—looks like a single kind of question; however, when it is replaced by a series of concrete questions, such as those we have just posed, several different kinds of questions are revealed. If we ask why a harvester ant explodes, we might expect an answer in terms of the environmental conditions, or stimuli, that led to the behavior, or of the neuronal impulses to its abdominal musculature that cause its cuticle to split and release the sticky contents of its mandibular gland. Either way, we are treating the ant as a machine, and asking what mechanisms it uses to produce its behavioral output. But another kind of answer might be returned to the same question. We might reply that the harvester ant explodes in order to defend its colony. This answer is of an independent kind, because its truth does not depend on any particular idea about the exact neuromuscular mechanism used to detonate the explosion; and, vice versa, the analysis of the mechanism is independent of what function the behavior performs. We can work out the mechanism whether or not we have identified the function correctly, whether the contents of the mandibular gland are a sticky defensive chemical (which in fact they are) or something else, such as food (which in fact they are not).

These two kinds of questions are sometimes called questions about "function" and "mechanism." Another pair of terms with the same meaning is the pair of "proximate causes" (i.e., mechanisms) and "ultimate causes" (function). An analysis in terms of neuronal control and environmental stimuli is an analysis of mechanisms, or of proximate causes; an analysis in terms of the advantage (such as to the ant colony) of performing the behavior is one in terms of ultimate causes, or function.

There are more than these two types of questions and answers. Consider next our questions about ducklings: how does a particular duckling come to follow one adult rather than another? The same two kinds of answers are again possible (in terms of bodily mechanisms and functional consequence), but the natural answer is now of developmental origin. Ducklings, as it happens, learn to follow whatever object they see near them during a "sensitive" phase after they hatch. That object is normally their mother, and different ducklings follow different adults because each duckling sees its own parent during that sensitive phase.

A fourth (and final) kind of answer to the general question of why animals behave is analogous to the developmental answer, but over a much longer time scale. In the question about ducklings, we explained the behavior in terms of lifetime experiences; we can do likewise for their different inherited, evolutionary ancestries. Thus, if we wish to explain why, say, the species of fish called three-spined sticklebacks (*Gasterosteus aculeatus*) court with their particular behavioral sequence (p. 200), but mallard ducks court with another quite different kind of behavior (pp. 26–27), we might say that sticklebacks are descended from ancestors that performed sticklebacklike courtship dances, and mallards from ancestors that performed mallardlike courtship, and that each of the modern forms has inherited the habits of their different ancestors.

The scientific method requires us actively to ask questions about the natural world, and to try to answer them. The kind of experimental work needed will depend on the question that has been posed. In the case of animal behavior, modern students of the subject customarily distinguish four kinds of questions and answers, rather than a simple general one, from the question of why animals behave. (The analysis into four questions was most famously made by Niko Tinbergen, though the relevant distinctions also had been made by Aristotle.) The four are questions about mechanisms, individual development, function, and evolutionary ancestry. All four can be asked about any piece of behavior, and we understand behavior most deeply when we know the answers to as many of the questions as possible.

The four questions are separate in the sense that they are independent of, but compatible with, each other: independent because any given answer to one question does not imply any particular answer to any of the other three, and compatible because all four can in principle be asked of any unit of behavior. The form of a question may invite one of the four kinds of answers—as we saw when asking why different ducklings follow different adult ducks—but the behavior can still be analyzed in the other three ways. We can still ask which mechanisms the duckling uses to distinguish adults, to establish its own preference for one particular adult, and to translate the perceptual distinction and learned preference into action; we can still ask about the ancestral history of filial responses in ducks, and birds in general, and their advantages or purposes in the lives of those birds.

It is worthwhile to distinguish the different kinds of questions, for it is a fact of history that students of animal behavior have frequently argued at cross purposes by confusing them. Questions about mechanisms and functions have particularly often been confused, as have questions about function and development. However, the distinction is a simple one, and I have established it at the outset mainly because it provides a convenient structure for thinking about animal behavior as a whole and in its parts. In Chapter 2, we shall consider the "ultimate" questions about function and evolutionary ancestry; Chapter 3 will discuss the question of mechanism; and Chapter 4 will discuss behavioral development. Then, equipped with the fundamental concepts, we can move on in Chapters 5–10 to apply them to the main kinds of behavior, as animals maintain themselves against their environments and against the society of other members of their own species.

However, before we start to consider the four questions in detail, we should discuss one other fundamental point. What is the nature of behavior? How can it be defined and recognized? If different observers do not agree on what behavior is, the scientific study of behavior will be impossible.

1.2   What is behavior?

The simplest definition of behavior is movement, whether it is the movement of legs in walking, wings in flying, or heads in feeding. But some actions of animals, such as the honking of peacocks, which we should wish to count as behavior, are not movements of the whole animal in the ordinary sense. The honking sound is produced as air is forced, by the contraction of muscles, out of the peacock's lungs, which causes a region of the throat to vibrate. There is movement here of the pulmonary musculature, just as there is muscular movement when an animal feeds or walks; in a more accurate sense, therefore, animal behavior consists of a series of muscular contractions. The wide importance of muscular contraction in many of our activities is reflected in the remark of the eminent neurophysiologist Charles Sherrington that "to move things is all that mankind can do; for such the sole executant is muscle, whether in whispering . . . or felling a forest." Some activities, such as the release of a pheromone by a female moth (p. 169) do not easily fit the definition of behavior as muscular contraction, but the definition is still a remarkably general one.

Naturalists had recorded incidental observations of behavior for many centuries, but no real attempt at the scientific study of behavior was made earlier than about a century ago. A crucial insight of the earliest workers—Charles Darwin, Oskar Heinroth, Konard Lorenz— was that behavior is orderly enough to allow that necessary criterion of all science, repeated observation. Behavior, or muscular contractions (they noticed), comes in orderly sequences, recognizable patterns of behavior that can be called behavioral "units." The same animal will produce the same pattern of movements again and again; different members of the same species will also behave in recognizably similar ways. Behavior can only be studied because of this fact. It makes it possible for an observer to check his own evidence, and for different observers to check each other's evidence. Without recognized units of behavior, anecdotes might accumulate, but each would be closed to criticism, and rigorous testing of theories would be impossible. Is it correct to assert that behavior is so regular? We can illustrate the principle by that classic example of a behavioral unit, the "egg-retrieval response" of the greylag goose. The greylag goose, which was Konrad Lorenz's favorite study animal, breeds in monogamous pairs. It nests on the ground, the nest being little more than an area of grass shaped into a bowl with the edge built up, though not to prevent an egg from occasionally rolling out. This is the occasion for the egg-retrieval response (Figure 1.4). When a goose sees an egg just outside its nest, it enacts the following sequence of muscular movements. Standing in the nest, it first extends its neck outward until its head is above the egg. It then puts the underside of its bill against the further side of the egg, and starts to roll it back. While rolling the egg, the goose moves its bill from side to side, to prevent the egg from slipping away to the side. The behavior is not always effective, and the egg may slip away. When it does, the goose does not immediately stop moving its bill backward and reestablish contact with the egg. Instead it moves its bill all the way back to the nest and only then, when it again sees an egg (in fact the same one) outside the nest, does it place its bill against the egg and try again. In other words, once started, the behavioral unit is continued until it is finished. Moreover, when Lorenz removed the egg from a goose while she was in the middle of rolling it back, the goose still continued and completed the sequence of movements. The two observations prove that sensory feedback of the feel of the egg against the bill is not needed to stimulate the continuing move-

Figure 1.4

Egg retrieval by greylag goose. The gull in Figure 3.12 (p. 66) is carrying out a similar task in an experiment. (*After Lorenz and Tinbergen*)

ment of the neck muscles. Behavior patterns can be recognized as units if they are performed often enough and in similar enough form. Of course, the egg-retrieval response of different geese, and of the same goose on different occasions, will not be exactly identical. But identical repetition is not necessary to define a unit of behavior. The behavior pattern on different occasions only needs to be sufficiently similar to be recognizable, and then the behavior units can be defined statistically. Behavior can be described inconsequentially as a series of movements, as we have just done for the egg-retrieval response of the greylag goose, but it can also be described in terms of its consequences. "Retrieve egg," for instance, is a consequential description; it does not mention the exact movements used, but does specify what results from them. For the general point being made here, it does not matter much which method we use to describe behavioral units; all that matters is that animals perform behavior patterns that can be recognized by different observers. Then scientific study becomes possible.

The point can be made another way. Darwin, Heinroth, Lorenz, and other early students of behavior would have been educated to think biologically about the anatomical parts of animals, rather than their behavior. They would have learned how to study parts such as limbs, urinogenital systems, and circulatory systems. When they came to think about behavior they naturally conceived and described behavior in the form of units, rather analogous to the limbs, kidneys, and hearts that can be seen in the anatomy of an animal. Feeding or courtship in animals can be studied in the same way as their anatomy and physiology. We could ask the same kinds of questions about them. Indeed, Heinroth and then Lorenz both started their work on animal behavior by applying biological methods, such as tracing the course of development of units of behavior in the life of an individual and looking at different species and trying to see the equivalent units of behavior in all of them. For it is not only the greylag goose that uses that recognizable egg-retrieval response. A similar sequence of muscular contractions is used by all other ground-nesting birds to retrieve their eggs, and other species of geese and all species of gulls show an egg-retrieval response that is recognizably the same unit of behavior. We shall not often meet the egg-retrieval response again in this book, as we shall be using other examples to illustrate the subject, but it is a classic example of a behavioral unit and has served its purpose by showing how it is possible to recognize sequences of muscular contractions as units of behavior. These objective units are necessary to make the scientific study of behavior possible.

1.3 Complex results can be produced by simple mechanisms: the spider's web

The observable behavioral output of an animal is a certain sequence of behavioral units. Although each unit is in itself simple (indeed they are defined to be simple), if combined they are capable of producing complex results. Consider the spider's web, for example (Figure 1.5). A human faced with the task of building such a geometrically complex structure would do so using a blueprint, or plan, of the web. They would then control their behavior by reference to their concept of the goal to be reached. Perhaps the araneid spiders, such as the common garden spider *Araneus diadematus*, which do build orb webs, have a concept of the web; we cannot tell. However, observations of spiders in action suggest that they follow a series of rules, which in themselves would be sufficient to lead the spider to build a web even if it had no idea of what the end point should be. Not all the flexible details of the spider's building program have been worked out, but the main rules were discovered simply by watching them build. The observations are easy to make. Orb webs are spun by common spiders all over the world, and the whole operation takes less than half an hour.

The end product, the orb web, is a regular structure made up of frame, radial spokes, and catching spiral (Figure 1.5). The stages of its construction were first formalized by Hans Peters in 1939, and they are as follows. The spider starts with the frame and spokes. To begin with, she spins a thread with one unattached end and allows it to be blown by the breeze. The other end is attached to the spider by the spinnerets of the abdomen, which are the source of the thread. The loose end soon becomes entangled in some object, such as a nearby twig. The spider then bites through her end of the thread (at A in Figure 1.6a) and, having

 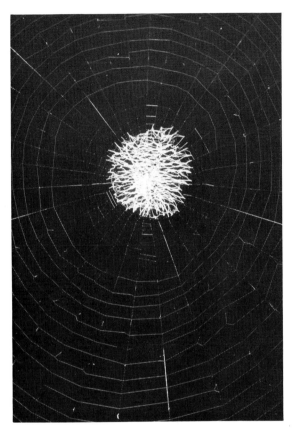

Figure 1.5   Orb webs are built by many species of spiders, such as (left) this British adult female of the common garden spider *Araneus diadematus* (the spider is in the center, at the hub of the web), and (right) an immature Floridan *Argiope aurantia*. *Argiope* spiders characteristically spin a "stabilimentum" using special silk at the hub of their webs. The stabilimentum is conspicuous, and has been argued by Tom Eisner to advertise the web to birds to stop them from flying through it unaware. The stabilimentum may also act to disguise the spider, as it does in this photograph. *(Photos: Fritz Vollrath)*

attached a new thread at her point of departure, walks off down the wind-cast thread, spinning another thread as she goes. When she arrives at the other end (B in Figure 1.6a) she attaches the new thread. She then turns, walks some distance up the thread (to point H in Figure 1.6b), and attaches a new thread there. She now allows herself to fall, spinning a thread behind her, to a third attachment point (C). The Y-shaped structure provides the scaffolding for the web. She next builds by turns the radial spokes and frame. She returns to the center H (which stands for hub, as the center of the completed web is called), attaches a thread, walks to the outside (B), spinning a thread as she goes, and attaches the other end of it at B. She then walks back down the thread (Figure 1.6c) to a point (D) where she attaches a new thread. Now, after she has walked on to A, she has constructed a frame from B to A and a radial spoke from D to H (Figure 1.6d,e). The same sequence of movements is reiterated to build the twenty or so radii of the web. In the final web the angles of the radii at the hub are rather constant, at about 15°. The radii are built in a regular order, the next spoke always being added in the largest unfilled sector of the web. In other words, the spider measures the angles between

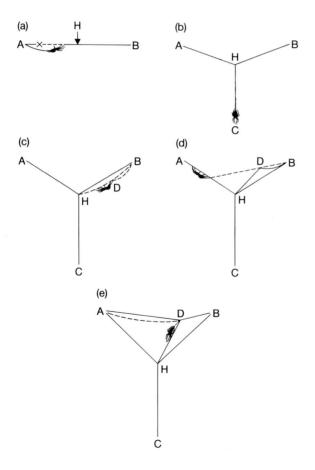

Figure 1.6

The stages in the construction of an orb web. See text for explanation. *(After Peters)*

the existing spokes at the hub and chooses to spin the next spoke between the two spokes subtending the largest angle. Once she has returned to the hub after building a spoke she pulls each spoke with her front legs, which may be how she measures the angles, perhaps by estimating the tension in each spoke. At all events, sight is unnecessary for correct building because spiders can build normal webs in complete darkness or when they are sightless. Following the rule for where to spin the next spoke results in spokes being laid down on the kind of order illustrated in Figure 1.7, with each new spoke in a different sector from the previous one.

Once the frame and radii are complete (which takes only about five minutes), the spider starts on the spiral. She builds it in two stages. Starting from the hub, she first spirals outward, to lay down an "auxiliary spiral." She does this by walking outward in an arithmetic spiral, spinning and attaching a thread behind her, until she reaches the edge. The auxiliary spiral is a temporary structure. For at the outside she turns around and, working inward now, spins the sticky spiral that will net her prey (Figure 1.8). As she spins the catching spiral she cuts the auxiliary

Figure 1.7
An orb web with most of its radial spokes but no catching spiral. The numbers indicate the order in which the radii are laid down. Notice that each radius is built in a different sector from the previous one.
(*After Peters*)

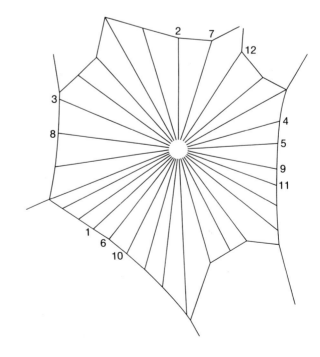

Figure 1.8
A partly completed orb web of *Araneus diadematus*. The auxiliary spiral is complete, but only a few spirals of the sticky catching spiral have been laid down. Notice that the catching spiral is much more tightly coiled than the temporary auxiliary spiral. *(Photo: Fritz Vollrath)*

spiral, the function of which was to act as a guideline for laying down the catching spiral. The whole web is finished once the sticky catching thread has spiralled into the hub. Actually, the spiral is not a perfect spiral; it is asymmetrical. Orb webs are not typically circular. They are elongated in the bottom half, with the hub above the center. The bottom is lengthened by means of a ladder of sticky switchbacks in the bottom sector of the web, which can be seen in Figure 1.5. Only after a few turns of the ladder does the spider begin the true spiral.

The main rules of the orb web spider are first to build a Y-shaped scaffold, then, in a set order, the frame and radial spokes, and finally the auxiliary and sticky spirals. Each main rule contains a set of subrules for measuring angles and walking set distances up certain threads. By following the program of elementary rules the spider can build a complex structure without having a plan of it in her head.

This is not to say that the spider lacks such a plan. Spiders surely are more complex machines than this account has admitted. There is probably more to the control of web building than the series of rules considered here. But this fact does not alter the point of principle which the spider's web illustrates: that apparently complex structures can be built up from a fairly simple set of rules. The principle is of general importance. It is often possible to break down the apparent complexity of animal behavior by analyzing it into a series of simple decision rules that the animal itself could, in some form, be following. Searching for the rules is a powerful method of studying the control of behavior, because it is easier to pursue research on simple rules than on complex ones. It is easier to follow up the nervous control of a leg movement than of conceptual thought. Of course, complex systems can be studied too, but because they are more difficult to study it takes longer. More sophisticated rules are most easily studied if they are added piecemeal on to a concrete foundation of knowledge. We therefore advance fastest in our understanding of behavior if we search for the simplest hypothetical mechanisms. Ethologists do not invoke less simple mechanisms—of conceptual thought or conscious calculation, for example—if they are not needed to explain the behavior that is being observed.

In summary, it is possible to understand a relatively complex piece of behavioral output by breaking it down into a sequence of behavioral units, or rules, that combine to produce the complex end result. As we shall see in Chapter 3, the units can be studied further because each rule must have some mechanism controlling when it is to be performed; but the point for now is that it can be useful to break down complex behavior patterns into sequences of simpler units.

1.4   Summary

1. The science of behavior is concerned with questions of why animals behave; these are a set of different kinds of such questions rather than a single one.

2. Four main kinds of questions and answers can be distinguished: By what mechanism is the behavior produced? What use is it to the animal? How did it develop? What is its evolutionary history?

3. The questions are compatible—they can all be asked of any particular behavior pattern. But they are independent—any particular answer to one of them implies little about the answers to the others.

4. The scientific study of behavior is made possible by the fact that the behavior of animals is performed in repeatable, publicly recognizable units.

5. Complex behavior patterns can be generated by a series of simpler behavioral units. The way spiders build webs is an example.

1.5   Further reading

Tinbergen (1963) originally distinguished the four kinds of questions about animal behavior. Barlow (1977) and M.S. Dawkins (1983) discuss how behavior is organized into units. Shear (1986) contains recent papers about spiders. Sparks (1982) is a popular history of the study of animal behavior, and Timberlake (1993) gives a recent perspective on the state of the science.

CHAPTER 2

The Evolution of

Behavior

In this chapter, we first consider evolution, and then adaptation and natural selection. In both cases, we consider the question in general first and then its application to animal behavior. We end with a section on how new bahavior patterns originate in evolution.

2.1   Evolution

Living things, as simple observation will reveal, come in distinguishable, recognizable forms such as robins, jays, and mockingbirds. These are what biologists call species. A species is a group of organisms that can breed with one another to produce another member of the same species; robins interbreed with robins, not with mockingbirds. All the offspring of robins are obviously robins; you could follow all the generations of robins produced during a human lifetime, and all of them would look like robins. This constancy of species is probably the reason why the idea of evolution only took root relatively recently. It is tempting to extrapolate from your own limited experience, and conclude that robins always have been descended from ancestral robins back through eternity. This leads to the theory—still believed by some—that every species is fixed, and has always had a separate ancestry from other species.

Most species indeed are constant in form over the temporal and spatial scale of normal human experience, but that apparent constancy disappears if the range of evidence is expanded. It is difficult to expand the time scale of evidence because the only obvious means is by examining fossils, and the fossil record is too poor to allow us to trace the ancestry, through a continuous series of fossils, of more than a few exceptional species. But geographical travel, to expand the scale of evidence in space, does break down the impression of the constancy of species. At any one place, species do appear as discrete groups of organisms, but if a species is traced across the world its appearance can usually be seen to change from place to place. House sparrows, for example, vary in size, body proportions, and coloration across the United States, and the house sparrows of North America visibly differ from those of Europe. At least all house sparrows are classified as house sparrows, but in sea gulls the range of geographic variation can break the species boundary. The herring gull and black-backed gull behave as two perfectly ordinary species in northern Europe, but they are connected by a continuous ring of intermediate forms around the North Pole. Geographical variation first led Darwin to doubt the constancy of species, particularly after he had returned from the circumglobal voyage of the *Beagle*. He had mixed his collections of finches from the various islands of the Galapagos archipelago because he thought they were all one form, but he soon realized that different species of finch inhabited the different islands. It was a striking discovery. How easy it now was to imagine that the different species of Galapagos finch had evolved from a common ancestor, rather than being separately created on their respective islands! For why, on the Galapagos, should the warblerlike birds and the woodpeckerlike birds, as well as the finchlike birds, all have been created as finches, when in the rest of the world ordinary warblers and woodpeckers arose (Figure 2.1)?

For this and other reasons, almost all biologists think that species are not fixed. They think instead that species change slowly, over long periods of time. Therefore, if we traced back the ancestors of robins, we should come to forms that, although recognizably birds, were not robins; as we traced the lineage to earlier and earlier ancestors we should

Figure 2.1
Most parts of the world are inhabited by some kind of woodpecker, such as, in Europe, the green woodpecker (*Picus viridis*) illustrated at the right here. On the Galapagos Islands a quite unrelated species of bird, a finch (*Camarhynchus pallidus,* illustrated on the left) has convergently evolved the same habit. It lacks a true woodpecker's tongue, and uses sticks to probe for insects.

come to forms that were more like amphibians than birds, and eventually to animals without backbones (invertebrates). Their very early ancestors were simple cells, which floated in the sea some 3500 million years ago. These cells, or perhaps some earlier form of life, were the common ancestor of all the species of animals and plants now alive. In any case, there is evidence to suggest that all living species share a single common ancestor. According to the theory of evolution, species have taken on their present appearance as they have changed from their ancestors.

The behavior of animals has presumably changed through evolution, just as their anatomical appearance has. However, for the hard parts of organisms we possess from fossils some evidence as to what they were like in the past, which is not available for behavior. Fossilized animals are necessarily dead; they do not behave, and although it is often possible to infer something about how a fossilized animal would have behaved, the inference is uncertain.

This does not mean we know nothing about the evolution of behavior. It means instead that our knowledge comes from another kind of evidence. In fact, the evolution of behavior is studied by comparing the behavior of different living species. Let us consider how this knowledge can be acquired. The first requirement is a phylogenetic tree, which shows the ancestral relations of modern forms (Figure 2.2). We still

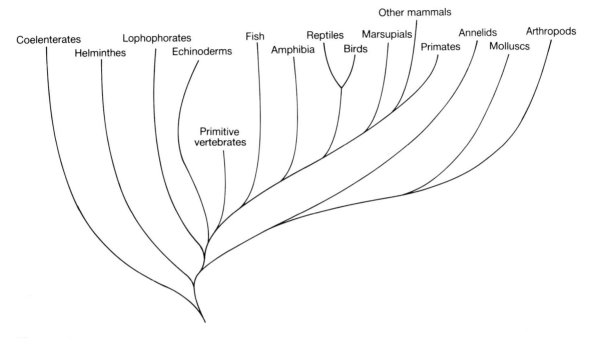

Figure 2.2   Phylogenetic tree of some large groups of animals.

know little about the phylogenetic relations of many animal groups, particularly of invertebrates; however, the general relations, and even the finer connections of some well-studied groups like birds and mammals, are reasonably well known. Phylogenetic trees like that in Figure 2.2 are mainly inferred from morphological and molecular evidence. Consider, for example, three of the groups of Figure 2.2: the birds, amphibians, and echinoderms. Two of them (birds and amphibians) resemble each other more closely in their molecular structure and their morphology (for example, they possess a hollow dorsal nerve chord, a backbone, and segmented muscles), than either does with the third group (the echinoderms, the group that includes starfish). We can infer that the birds and the amphibians share a more recent common ancestor with each other than either does with the echinoderms by the following argument. The common ancestor of all three probably lacked those characteristics (such as having a backbone) shared by birds and amphibians, because the common ancestor of all living things lacked them. Thus, if we put the birds and amphibians together in one group in the phylogeny and echinoderms in another, the characters like the backbone only need to have evolved once. If we had instead put the birds and echinoderms in one group and the amphibians in another, the backbone would have had to have evolved once in the amphibian lineage and another time in the bird lineage, making two evolutionary events. (There are other ways of reconstructing the evolutionary events in these two cases, but they all require more events if birds and amphibians are separated than if they are put together in one group.) We infer that the groupings of birds and amphibians on the one hand, and echinoderms on the other, is correct because it requires fewer evolutionary events; this principle is sometimes called "parsimony": we infer that the correct phylogeny from a number of alternatives is the one requiring the smallest number of evolutionary events. The exact detail of how phylogenies are reconstructed is not crucial here. The main points are that it can be done, and that most real phylogenies are reconstructed from morphological and molecular evidence. We can therefore use phylogenies like that in Figure 2.2 to study the evolution of behavior without the danger of falling into circular argument.

If we know the phylogeny of a group of species, we can infer whether a behavior pattern that is found in more than one group of these species has evolved independently in each of them (which is called convergent evolution) or evolved only once, and is now shared in the different groups by descent from a common ancestor. The rule is that if the different groups in question lie far apart in the phylogenetic tree and are separated by many groups with different habits, the behavior pattern is probably convergent.

The social insects contain striking examples of convergence. The social insects are the species that live in large colonies; there are two main groups of them, the termites and the social hymenopterans (the ants, bees, and wasps). Both lie within the enormous group of "Arthropods" in Figure 2.2, but there are a large number of other insect groups, which are not social in their habits, between the termites and the hymenopterans in the phylogeny. It is therefore more likely that the social habit evolved independently in the two than that the common ancestor of all insects lived in large societies and the habit has since been lost many times, in all the other insect groups. (Indeed, the habit of living in societies has probably evolved more than once in the Hymenoptera—it may have evolved independently over 11 times in that group alone.)

The ants and termites have evolved in some remarkably similar lines. For instance, species of both groups have independently evolved soldier castes, and in both cases the soldiers have evolved enlarged heads and mouthparts (Figure 2.3). There is even a termite species that has evolved the same habit as the exploding harvester ant. E.O. Wilson describes the soldiers of the termite species *Globitermes sulphureus* as:

> quite literally walking chemical bombs. Their reservoirs fill the anterior half of the abdomen. When attacking, they eject a large amount of yellow liquid through their mouths, which congeals in the air and often fatally entangles both the termites and their victims. The spray is evidently powered by contractions of the abdominal wall. Occasionally these contractions become so violent that the wall bursts, shedding defensive fluid in all directions.

The habits of the two species are similar (although there are some minor differences) but have certainly evolved convergently. If it were

Figure 2.3
The "soldier" castes of many species of ants and termites are an example of evolutionary convergence; they have evolved independently in the two groups. The soldiers in both the species illustrated here have enlarged mouthparts for gripping and crushing. The ant species is *Pheidole tepicana* (left), the termite species *Prorhinotermes simplex* (right). Soldiers of other termite species possess a "nasus," a chemical spray gun, on their heads. The form of the soldiers' adaptations varies among species, but in all cases they are adapted to defend their nest mates from live enemies. (*After Wheeler and Banks & Snyder*)

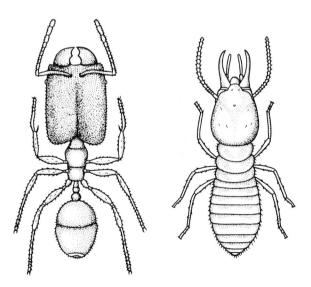

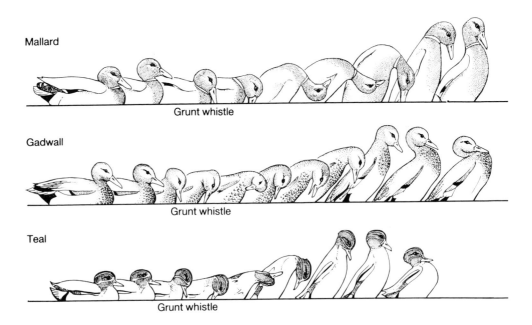

Figure 2.4   The courtships of male mallards (top), gadwall (middle), and teal (bottom) are made up of sequences of distinct displays. Konrad Lorenz observed the courtship of different duck species and compared the displays. The comparison among species reveals the nature, if not the direction, of their evolutionary modification. *(From Lorenz K. The evolution of behavior. © 1958, Scientific American, Inc.)*

not convergent the common ancestor of termites and ants would have had the explosive defensive habit; so too then would the continuous series of thousands of species (of which there are hardly any remains) connecting the modern ants with the modern termites. Every species of insect phylogenetically intermediate between ants and termites would then have been descended from a common ancestor with the explosive habit, but in all those species the habit must have been lost as no other living insects except *Camponotus saundersi* and *Globitermes sulphureus* are known to explode in this way. We should thus have to infer a large number—perhaps many thousands—of independent losses of the habit. Whereas, if the habit is convergent in the two species, we do not need to invoke any (now unknown) intermediate ancestors with the

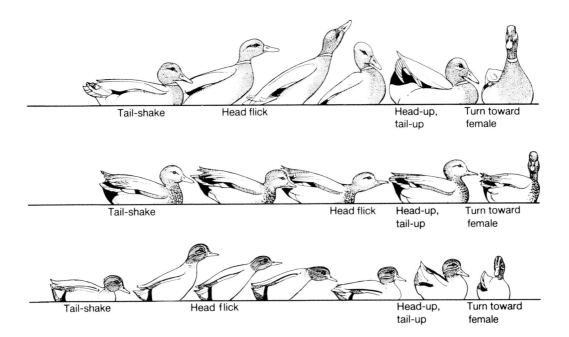

Tail-shake   Head flick   Head-up,   Turn toward
                                    tail-up   female

Tail-shake   Head flick Head-up,   Turn toward
                                tail-up   female

Tail-shake   Head flick   Head-up,   Turn toward
                                    tail-up   female

habit, nor all those thousands of evolutionary losses. It is therefore simpler and more parsimonious to suggest convergence.

So much, for now, for convergence. If we are studying a habit shared by a group of closely related species we can carry out another kind of analysis. The courtships of ducks are a good example. They were first studied systematically by Heinroth, and later in a classic study by Lorenz. Lorenz observed the pattern of courtship and other kinds of social behavior in 20 species of ducks and geese. He divided the behavioral repertory of each species into behavioral units. Some of the habits are only found in a few species, but others, such as "monosyllabic lost piping" (a distress call of the chicks), were found in all species. Figure 2.4 illustrates how the same unit may be recognized in different species: the teal, gadwall, and mallard, for example, all perform a similar sequence of movements (and sounds) that Lorenz calls the "grunt-whistle" display. Once this has been recognized, two kinds of inference become possible. One is to infer the evolutionary changes undergone by the behavioral unit. We can see, for instance, how the "headflick" display differs among three species in the figure; to take one feature, the gadwall appears to perform this display nearer the water than does the mallard. Any behavioral unit, although it may be recognizable in more than one species, will not be exactly the same in all of them. Just as anatomical units, such as the limb bones, can be traced from the fins of fish to their very different form in the legs of birds and reptiles, so can the evolutionary changes of behavior be traced by comparison among species. It is sometimes difficult—or even impossible—to discover the direction of the changes through evolutionary time, but inferences can be made by the same techniques as are used with anatomical evolution. In either case, the nature, if not the direction, of change can be worked out.

The second kind of inference puts into reverse an earlier statement. I said that our phylogenetic knowledge is mainly derived from anatomical and molecular evidence; this is true, but there is no reason why behavioral evidence should not be used in the same way. If the phylogenetic relations of species are suggested by the similarity of their morphological appearances or the sequences of their proteins, they can also be suggested by the similarity of their behavior. Lorenz used this fact in his work on the behavior of ducks and geese. He had recorded a total of 48 different behavioral units in 20 species. He could then group the species according to how similar they were with respect to these units. In Figure 2.5 the vertical lines represent the 20 species, which have been arranged according to shared behavior patterns (indicated by the horizontal lines). If we thought that behavioral similarity between two species implied a recently shared evolutionary ancestry, then we could infer from Figure 2.5 that, for instance, the pintail and the mallard share a more recent common ancestor than either share with the shelduck. The grouping of the species according to their behavior is similar to, but not exactly identical with, the groupings that have been suggested by other biologists who use evidence from anatomical similarity; it is possible to investigate why the different kinds of evidence suggest slightly different phylogenetic groupings. The main point here is that Lorenz's phylogenetic work on the courtship of ducks illustrates how we can study the evolution of behavior. Behavioral units can be compared between a number of species, and we can see which species share which behavior patterns and how those behavior patterns have changed in form in related species.

2.2   Natural selection and adaptation

After we have accepted that behavior has evolved—that it has changed over evolutionary time—the next question is why it has evolved through the course that it has. The question is a special case of the more general question of why evolution occurs at all. The generally accepted answer, for the majority of—if not all—evolutionary change, is the principle discovered by Darwin and called by him *natural selection*. If we first establish what natural selection is, it will become clear why it drives evolution. And we can then consider the importance of the principle for animal behavior.

Natural selection is easily understood. Its operation follows logically from a series of elementary propositions. The first is that individuals within a species differ from one another; they show variation. If you were to measure some trait in many members of a species, for instance the feather color of mallard ducks or the rate at which the "grunt-whistle" display is performed, the trait would almost certainly show variation—almost every trait that has been studied, in any species, anywhere, has been found to vary among individuals. The second proposition is that the variation is often, to some extent, inherited. Mallards with brighter than average feathers should tend to produce offspring with brighter than average feathers (although, so far as I know, this particular trait has never been studied). The third proposition is that the members of all species produce very many more offspring than can ever survive. At one extreme, female cod produce millions of eggs, but even the larger mammals, which produce only one offspring every year or two, still produce many more offspring than could survive. Darwin calculated, for instance, that a single pair of elephants could have 19 million descendants alive about 750 years after their birth. Clearly, whether in cod or elephant, most of the offspring must die. If any inherited variant of the species is, in the slightest degree, more likely to survive to reproduce, then evolution by natural selection must operate. That

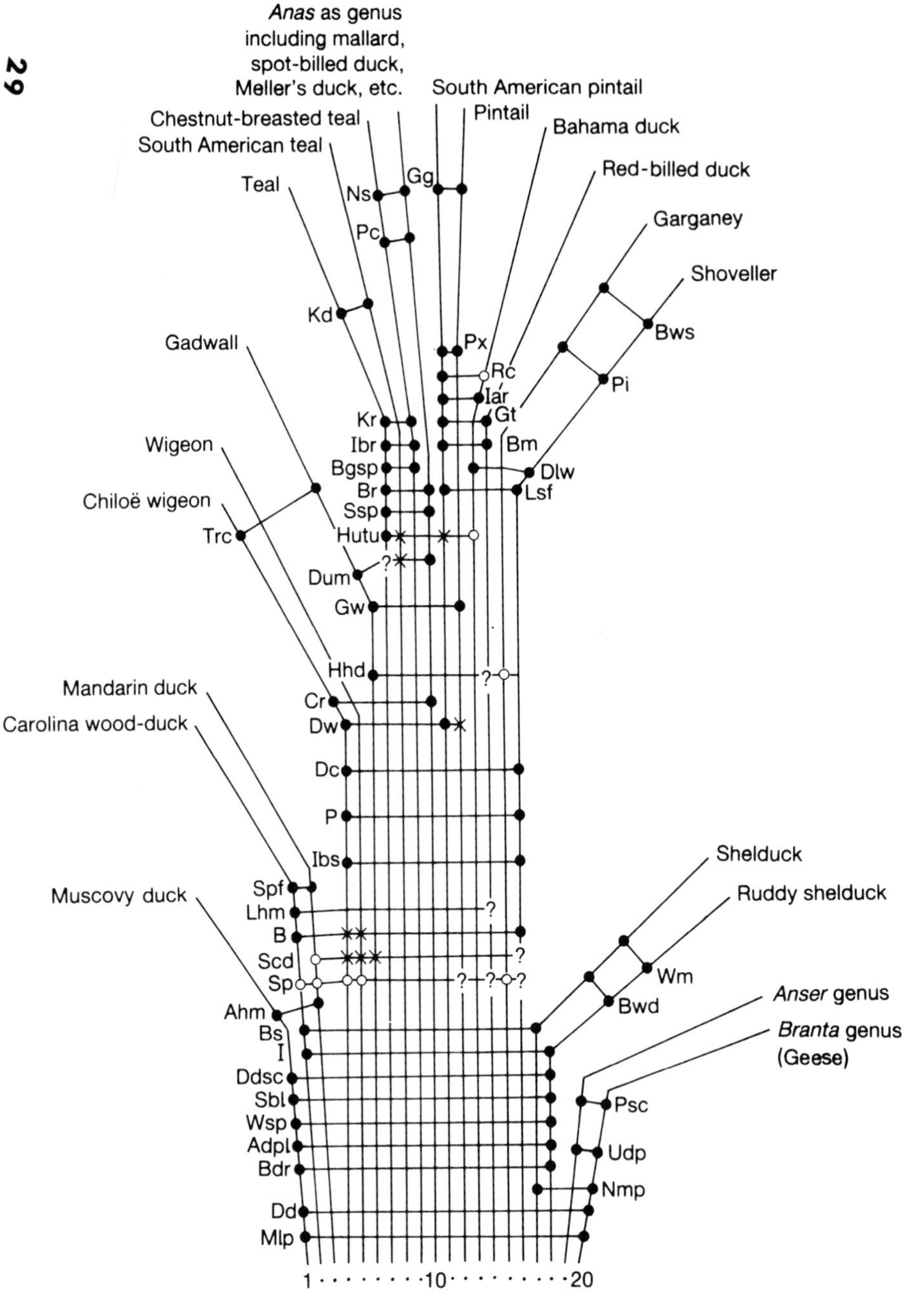

Figure 2.5   Shared behavioral patterns can be used to classify species in the same way as morphological traits. The figure shows Lorenz's classification of 20 species of ducks by their similarities with respect to 48 behavioral traits. The names of the behavioral traits are:

Mlp monosyllabic "lost-piping"
Dd display drinking
Bdr bony drum on the drake's trachea
Adpl Anatine duckling plumage
Wsp wing speculum
Sbl sieve bill with horny lamellae
Ddsc disyllabic duckling social contact call
I incitement by the female
BB body-shaking as a courtship or demonstrative gesture
Ahm aiming head-movements as a mating prelude
Sp sham-preening of the drake, performed behind the wings
Scd social courtship of the drakes
B "burping"
Lhm lateral head movement of the inciting female
Spf specific feather specializations serving sham-preening
Ibs introductory body-shaking
P pumping as prelude to mating
Dc decrescendo call of the female
Br bridling
Cr chin-raising
Hhd hind-head display of the drake
Gw grunt-whistle
Dum down-up movement
Hutu head-up-tail-up
Ssp speculum same in both sexes
Wm black-and-white and red-brown wing markings of Casarcinae

Bgsp black-gold-green teal speculum
Trc chin-raising reminiscent of the triumph ceremony
Ibr isolated bridling not coupled to head-up-tail-up
Kr "krick"-whistle
Kd "koo-dick" of the true teals
Pc post-copulatory play with bridling and nod-swimming
Ns nod-swimming by the female
Gg *Geeeeegeeeee*-call of the true pintail drakes
Px Pintail-like extension of the median tail-feathers
Rc R-calls of the female in incitement and as social contact call
Lar incitement with anterior of body raised
Gt graduated tail
Bm bill markings with spot and light-colored sides
Dlw drake lacks whistle
Lsf lancet-shaped shoulder-feathers
Bws blue wing secondaries
Pi pumping as incitement
Dw drake whistle
Bwd black-and-white duckling plumage
Psc polysyllabic gosling social contact call of Anserinae
Udp uniform duckling plumage
Nmp neck-dipping as mating prelude
(After Lorenz)

variant will survive better, leave (on average) more offspring (which will resemble their parents), and therefore increase in frequency in the population. The next generation will contain more individuals like it than the previous generation did—the population will have slightly altered and evolution will have taken place. If the conditions are maintained, the variant that survives better may continue to increase in frequency until it makes up the entire population. (The survival and reproductive success of a kind of organism is called its *fitness*; if one variant leaves more offspring than another, it is said to be "fitter." This use of the word "fitness" has no relation to its athletic meaning.)

Natural selection has been observed in operation in many cases, of which by far the clearest example is the research of H.B.D. Kettlewell on the peppered moth *Biston betularia* in Britain. In this famous study there were two main variant types of the moth: a lighter peppered colored type and a darker ("melanic") type. The difference is inherited; that is, the offspring of the dark type are more likely to be dark than are the offspring of the light type. The moths are eaten by birds that hunt their prey by sight; each type survives better (because it is less vulnerable to predation) against a different background. The light type survives better on lichen-covered trees, and the dark type on the dark, lichen-less trees of industrially polluted areas (Figure 2.6). When the proportions of the two kinds of tree are altered, natural selection causes a change in the proportions of the two types of moth. When, in particular, the proportion of dark trees increased in Britain after the early nineteenth century industrial revolution, the dark type of moth increased from a rare minority to be the majority form within about 50 years.

There are two conditions for natural selection to drive evolutionary change: the environment itself must change to alter the advantages of the different types, and the difference must be inherited. If the environment does not change, natural selection will probably maintain the species in constant form. Only when the environment changes is natural selection likely to cause evolutionary change; even then it can only do so if some kinds of individuals of the species survive to reproduce better than others, and if the ability that enables them to survive better is inherited. If the difference between types is not inherited, no evolution will take place even

Figure 2.6

H.B.D. Kettlewell (below) placed samples of the two types of peppered moth on trees in polluted and unpolluted areas of England. By observing from a nearby hide, he measured the rates at which birds took the two types in different places. Birds, such as robins and redstarts (above) took the more conspicuous types at higher rates. Tinbergen collaborated in this work in the 1950s. Peppered moths are now known not to settle on tree trunks, but in the upper twigs and branches, but this does not affect the main conclusions of the research. *(Photo: Niko Tinbergen; predatory birds redrawn from photos by Niko Tinbergen)*

if one type does survive better than another. It could be, for instance, that the bigger members of a species survived better than the smaller ones, but size might not be inherited: that is, bigger than average parents might not leave bigger than average offspring. (Differences in size might be controlled by noninherited accidents, such as how much food an individual happened to find.) In this case, the average size of the species would stay constant even though bigger animals were surviving better. Only if the size of an individual became an inherited property could evolutionary change toward larger size take place. If any trait is to evolve under natural selection, differences in the trait must be inherited. (Chapter 4 describes the mechanism of inheritance, particularly for behavior.)

Natural selection is the reason why species evolve, but its importance for the student of animal behavior lies less in its explanation of evolution than of *adaptation*. Adaptation, in biology, refers to the fit between an organism and its environment, to the adjustment of the parts of the organism to the needs of its way of life. (It is a static concept, referring to the "designful" aspects of an organism at any time. The biological use of the term therefore differs from the nontechnical meaning in which "adaptation" refers to the way an individual may change over its lifetime to fit itself to the demands of life.) Camouflage is a particularly clear example of adaptation. Peppered moths are less likely to be eaten by birds if they are camouflaged against the forest background in which they live; their color pattern is adaptive. The moth's behavior must also be appropriately adaptive if the camouflage is to work. The moth has to recognize the correct setting to settle on and adjust its movements and orientation to the background. (Moths indeed have this ability: see Chapter 6, section 6.6.) The behavior of animals is adapted just as much as their anatomy and physiology. However, it is not always obvious how particular habits are adapted to the way of life of the organism; many of the most interesting studies of animal behavior have sought to uncover the advantage of particular habits—be they self-explosion in harvester ants, or extravagant courtship displays in birds of paradise.

Adaptation is so common and striking a property of living things that its existence cries out for explanation. Adaptation could not have arisen by chance (indeed, adaptation is almost defined to rule out chance as its explanation). When natural selection operates, it brings adaptation into existence. As birds eat the less well-camouflaged moths, the better camouflaged moths increase in frequency in the population: that is why we now see well-camouflaged moths. In ancient moth populations, poorly camouflaged moths did not survive to become ancestors of the moths we now observe. As the explanation for adaptation, natural selection is the key to one of our four questions—the question of function, or ultimate causation—about why animals behave.

Not every conceivable kind of advantageous trait can evolve under the power of natural selection. Natural selection works by differential reproduction. To a first approximation, it can only favor traits that increase the number of offspring left by the organism. When we ask why an animal performs a particular behavior pattern, if we are to return an accurate answer it is necessary to translate the question into "how does that behavior enable that kind of organism to produce more offspring?" or "why would an animal that performed a different behavior pattern leave fewer offspring?" If a postulated advantage will not translate into more offspring it cannot be the true explanation of the behavior. This kind of inquiry is usually a matter of finding out to what particular need or property of the species' lifestyle and environment a behavior pattern is adapted. In the case of the peppered moth the answer is obvious. The adaptation is the color pattern, and the properties of the environment to which it is adapted are the tree branches it rests on and the birds that eat the moths. The reason why a moth of some other color pattern would leave fewer offspring than the camouflaged type is, quite simply, that it would be more likely to be eaten before it reproduced. In some of the behavioral adaptations we shall come to, particularly those concerning social behavior, the advantages are less obvious. But we should establish the principle by an unambiguous example to begin with; it is then easier to apply in the less obvious cases.

2.3   The evolution of new behavior patterns

How does a new behavior pattern arise in a species' behavioral repertory? There is probably more than one route, and we shall consider only one of them—possibly the most important—here. To find an example, we can turn to the other classic ethological study of behavioral evolution (Lorenz's work on ducks was the first): Tinbergen's study of gulls. In the 1950s and 1960s, Tinbergen and a series of collaborators and students investigated the behavior of the various gull species around the coast of the United Kingdom. One of these species, the kittiwake (*Rissa tridactyla,* studied by Esther Cullen) had particularly distinctive behavior. Kittiwakes have the peculiar property of nesting on narrow cliff ledges, unlike most other gulls, who nest on spacious dunes back from the coast.

Kittiwakes have a number of distinct behavior patterns associated with their breeding habits. One example is a behavior pattern of kittiwake chicks; it is called the *cliff edge avoidance response,* and it was experimented on by Heather McLannahan. When a chick sees the edge of the cliff, it goes through a series of movements in which it stops, turns around, and walks off in the other direction. It is possible to test whether chicks of other gull species have this ability by putting them on a "visual cliff" (Figure 2.7); on this apparatus, the kittiwake chick turns around at the apparent cliff edge like at a natural edge. Chicks of other species either show no cliff edge avoidance response—they just march over the apparent edge as if nothing were there—or a less well-developed one than in kittiwakes.

The kittiwake's cliff edge avoidance response is an adaptation and evolved by natural selection in the evolutionary lineage from ancestral gulls to modern kittiwakes. But how did it evolve? The most plausible answer is, by combining preexisting perceptual and locomotory abilities. The behavior pattern did not suddenly appear as a whole; it originated as a small modification of the ancestral gulls' behavior. The ancestral gull chick already had eyes and the perceptual ability to recognize a cliff edge, even though that sensory perception did not stimulate it to behave in a special way. Moreover, the ancestral gull chick already had legs and a locomotory system that enabled it to turn around and retreat, even though it did not particularly tend to do this when its visual system perceived a cliff edge. All that was needed for the evolution of the cliff edge avoidance response was for the existing perception of a cliff edge to

Figure 2.7
A kittiwake chick on an experimental apparatus designed to test the chick's cliff edge avoidance response. The chick is just turning back from the transparent part of the apparatus, at the visual "cliff-edge." The experiments were done by Heather McLannahan, who is here watching the chick. *(Photo: Niko Tinbergen)*

become combined with the locomotory habit of turning around and walking back; they would then combine to form a new (and adaptive) behavior. It presumably first appeared in one chick in the ancestral lineage leading to modern kittiwakes, and then spread through the population by natural selection.

Therefore, it requires no remarkable changes to produce new behavior patterns. New behavior patterns can arise by combining preexisting behavioral units in novel forms. Although not all behavioral evolution may be by this means, it has probably been common. As we come to consider in later chapters, the evolution of various other behavior patterns—such as signals, or altruism, for instance—we shall see how new signals may have been derived from existing, nonsignaling behavioral units, and how altruistic behavior (such as a bird feeding its sibling) may have evolved from the existing behavior patterns of parental care.

2.4   Summary

1. All modern species have descended from a common ancestral species. During the evolutionary changes of species from their ancestral forms, their behavior must have changed as well.

2. Evolutionary changes in behavior are studied more by the comparison of modern forms, using our knowledge of phylogeny, than by studying fossils. The course of behavioral evolution can be inferred by comparing the behavior patterns of species known to be descended from a recent common ancestor.

3. Similar behavior in different groups of species is inferred to be convergent if the groups are phylogenetically distant and separated by species showing different behavior.

4. Evolution has taken place because of the process of natural selection. Natural selection favors those types of animals that leave more offspring than the average for their species.

5. Natural selection is the reason why the behavior of animals is adapted to their environments. To understand the behavior of animals we need to find out how it enables them to reproduce more than they otherwise would.

6. New behavior patterns evolve by modifying existing behavior patterns, and by joining together previously separate behavior patterns. Behavioral evolution is in this sense gradual.

2.5   Further reading

Ridley (1993) is a modern text about evolution. Darwin's *Origin of Species* remains an interesting read; I recommend the first edition (1859—several reprints are available). There are many good modern accounts of adaptation and natural selection: e.g., Dawkins (1986, 1989). Reeve and Sherman (1993) and Williams (1992) contain more advanced reflections on the subject. Wilson (1971) is a magnificent account of the social insects, which includes a discussion of the soldier caste, and is the source of my quotation. Lorenz (1958) describes his work on the behavior of ducks. On kittiwakes, and the gull study in general, the best source is Tinbergen (1974). Cullen (1958) and McLannahan (1974) are the primary references. Tinbergen also made a film about gull behavior; it is called *Signals for Survival* and is relevant to this chapter although its main theme is signaling (the topic of Chapter 7).

CHAPTER 3

The Machinery of

Behavior

In this chapter we shall be treating animals as machines, and seeking to understand how they work. The nervous system is the main means of controlling behavior, and we begin by looking at how information is transmitted among the nerve cells. Information about the outside world enters the animal's nervous system through its sense organs. We consider sensory systems both in general and in one detailed example, the sonar system of bats. Much of the behavior of animals is neurophysiologically complicated, and we shall consider some examples of how animal senses have been successfully studied at a higher level than that of sensory neurons, and their behavioral choices at a higher level than the influence of neurons on one another. Finally, we consider how hormones influence behavior.

3.1   The nervous system

3.1.1 Neurons transmit information by spikes of electrical depolarization

Branching out throughout the body of an animal is a network of thin white fibers called nerves. A nerve fiber is a bundle of cells called neu-

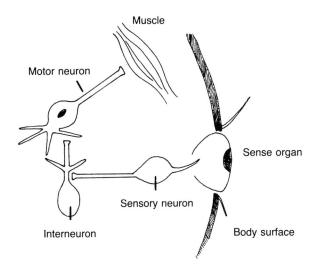

Muscle

Motor neuron

Sense organ

Sensory neuron

Interneuron

Body surface

Figure 3.1
Diagram of the relations between the three main kinds of nerve cells.

rons. We can distinguish three main kinds of neurons; and the actions and interactions of the three control the body's behavior. The three kinds are motor neurons, interneurons, and sensory neurons (Figure 3.1). Sensory neurons, which are discussed in the next section, encode information from the body's sense organs. There are many "orders" of interneurons between the sensory and motor neurons: first order interneurons are attached to sensory neurons, then second order interneurons are attached to those first order interneurons, and so on through increasingly higher order interneurons. Interneurons carry out all the complex information processing activities of the nervous system, and the brain consists almost entirely of interneurons. Interneurons extract information from the raw sense data of the sensory neurons, integrate the information of different sensory and internal systems, and then (if appropriate) issue instructions to the third class of neurons, the neurons that most directly control behavior—the motor neurons. Motor neurons attach to muscles, and when enough of the motor neurons attached to any one muscle become electrically active, the muscle contracts and some externally visible movement may result.

Neurophysiologists first worked out how neurons work in one particular kind of neuron that is found in the squid. When a squid senses that an enemy is nearby, it suddenly contracts its body, forcing water out of the area called its mantle cavity, and jets away. The squid may also discharge a screen of ink to cover its escape. This "escape reaction," as it is called, is controlled by a large (and therefore easily studied) neuron (Figure 3.2), which extends in stages from just outside the squid's brain at one end of its body, to the muscles, the contraction of which at the other end of the body effects the escape reaction. The nerves controlling the escape reaction are an example of what is called a "dedicated system": these nerves are "dedicated" to the escape response, and in general a dedicated system is a system of nerves that control one behavior pattern. Most progress has been made in studying the nervous control of behavior in this kind of dedicated system, because it is relatively easy to correlate the activity of certain neurons and the behavior of the animal. However, by no means are all of an animal's nervous subsystems "dedicated"; many are multipurpose—they are neurons that can function in many different behavior patterns. It is more difficult to correlate neuronal activity and observable behavior in a flexible, multipurpose control system.

Microelectrodes stuck inside active neurons revealed that the neuron works electrically. A neuron, like any cell, consists of certain characteristic internal contents surrounded by a membrane. In its normal, "resting" state a neuron has a small electrical field across its membrane. Microelectrodes reveal that the inside of the cell has a small negative electric charge relative to the outside. If the neuron is then given a small electric shock, the properties of the membrane rapidly change at the site of the

Figure 3.2   The course of the giant neurons of the squid. The escape reaction is effected by the retractor muscle, and the giant fibers stimulate it medially and posteriorly. (*After Bullock and Horridge*)

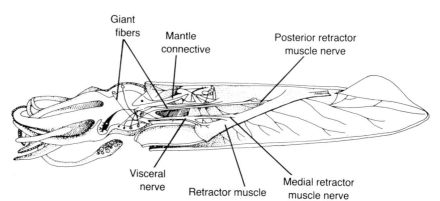

Giant fibers

Mantle connective

Posterior retractor muscle nerve

Visceral nerve

Retractor muscle

Medial retractor muscle nerve

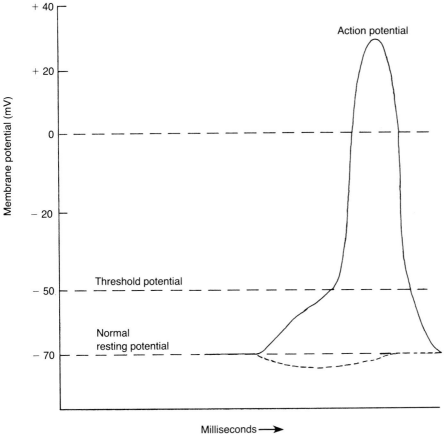

Figure 3.3

Depolarization of a neuron. The normal resting potential of the neuronal membrane is about −70 mV. When depolarized, positively charged sodium ions enter the cell, changing the membrane potential to about +30 mV. The sodium ions are then pumped out and the original resting potential restored.

shock, such that positively charged particles (sodium ions, in particular) now rush into the cell. The electric charge across the membrane changes from negative to positive; the event is called "depolarization" (Figure 3.3). This local depolarization acts as a stimulus to the neighboring region of the neuron, which responds in the same way: the membrane alters, sodium ions pour into the cell, and the electric charge across the membrane goes positive. This, in turn, stimulates the neuron a bit farther down, and a wave of depolarization shoots down the neuron from the site of the original stimulus. Once the neuron has been stimulated, it becomes self-stimulating down all its length. After the depolarization, the membrane reverts to its original state, the sodium ions are pumped out of the cell, and the resting state with its negative electric charge is restored. The membrane is then ready to conduct the next wave of depolarization, when it comes down the cell.

3.1.2 Neurons interact at synapses

The events we have considered followed an initial electrical stimulus, such as one given by an experimenter. What starts the wave of depolarization in a natural neuron? Neurons are, structurally, "directional": a motor neuron, for example, is attached to a muscle at one end and to an interneuron at the other. The muscles can never stimulate the motor neuron, but the interneuron can. The events by which the interneuron initiates depolarization take place at the junction between the two neurons; the junction is called a "synapse." When an electrical depolarization has traveled down an interneuron, it arrives at the synapse and there sets in train a new set of events. The membrane at the end of the neuron alters, under the influence of the depolarization, and releases a kind of chemical called a neurotransmitter. There is a small space at the synapse between the two neurons and the neurotransmitter diffuses into this space. The neurotransmitter, in turn, binds to special receptors on the membrane of the neuron on the other side of the synapse, and the compound of neurotransmitter and receptor stimulates (to some small extent) the neuron. If a sufficient amount of neurotransmitter stimulates the neuron, it will burst into action and conduct a wave of depolarization down to its other end. Thus the answer to the question of what stimulates depolarization in a neuron is that it is in turn stimulated by another neuron (or set of neurons) prior to it in the nervous system.

Neurotransmitters are a behaviorally important class of chemicals in their own right. Two common neurotransmitters are acetylcholine (ACH), which is an excitatory neurotransmitter in many motor neuronal systems, and γ-acetylbutyrine (GABA), which is a common inhibitory neurotransmitter. Many other neurotransmitters are used in specialized neuronal subsystems. The neurons in our brains responsible for our feelings of pleasure and pain have receptors (called opioid receptors) that bind molecules called opioids. The best known opioids are the β-endorphins, which are the neurotransmitters that are best known for causing the runner's "high" sensation, but generally function as pain blockers when the body is under serious assault. The opiates heroin and morphine are chemically similar enough to β-endorphin that they bind opioid receptors, and suppress pain. Many other psychoactive drugs are either known or suspected to mimic natural neurotransmitters. Indeed, artificial psychoactive substances can be important clues in research that aims to discover previously unidentified neurotransmitters. Cannabis (or marijuana) provided the clue in the most recent discovery of this kind. The active compound in cannabis is a chemical called tetrahydro-cannabinol (THC), and THC is known to bind receptors in various parts of the brain. In 1993, the natural analogue of THC was identified: it is a previously unknown brain chemical called anandamide. The natural function of anandamide has yet to be described, but given the effects of cannabis, it probably influences nerve circuits involved with pain and with memory. Similarly, many neurophysiologists suspect there is a neurotransmitter similar to the drug lysergic acid diethylamide (LSD), but (if it does exist) it has yet to be found.

3.1.3 Neurons transmit information in digital form

A single neuron can convey only limited information, because the neuron is either "on" (if it is transmitting a depolarization) or "off." Neuronal information is therefore "digital": it has a $+/-$ or on/off form. Different degrees, or grades, of information can be conveyed by the frequency of depolarizations, but they are not conveyed by different degrees of depolarization. If a neuron is stimulated strongly, it will burst with depolarizations at high frequency, but it will not convey a particularly large depolarization. The reason is that the degree of electrical change in a depolarization is set by a special chemical system: when the cell is depolarized, its membrane changes in a fixed way, letting in a fixed quantity of sodium ions and producing a change of $+100$ mV, after which the membrane automatically alters back. The reason neurons use a digital information system is probably to prevent information decay. If a neuron is switched "on" at one end, the depolarization passes down what can be a considerable distance (often several centimeters, and up to several meters in a neuron passing up and down the camel's neck), and the signal "neuron active" reliably reaches the other end. If instead there had been a graded signal (such as "neuron switched on to 70% of maximum"), by the time it passed to the other end it might be read as (for instance) "50% on" or "90% on"; once that kind of error has passed through several units in a system, the signal rapidly decays into insignificance. (By the way, some neurons in arthropods do transmit "graded" electrical signals. They are all short neurons, which supports the idea that most neurons use digital signals because of the problem of information decay over a distance or in complex networks.)

3.1.4 Information can be transferred through synapses in many ways

Information processing that uses graded signals takes place between neurons at the synapses, which enables the flexible control of behavior. Information transfer down neurons is simple, but information transfer at synapses is much more flexible. There are two main kinds of synapse: excitatory and inhibitory. Interactions between any two neurons are always either excitatory or inhibitory (or some combination of the two). Any one neuron may have any number of other neurons forming synapses with it. Some of the synapses may be inhibitory, others excitatory. For simplicity we can consider a neuron that has either two excitatory synapses, or one excitatory and one inhibitory synapse (Figure 3.4). The two excitatory synapses could act additively: if one of the presynaptic neurons is active it may not by itself be sufficient to make the postsynaptic neuron fire, but if the other presynaptic neuron is also active, it can build on the effect of the first, and the combination of the two makes the postsynaptic neuron fire. With an inhibitory and excitatory synapse, if the presynaptic neuron with the inhibitory synapse fires, it makes the postsynaptic neuron less likely to fire should the excitatory synapse become active. It is possible to imagine that a neuron connected to a large number of other neurons, having various inhibitory or excitatory influences on it, will be able to show complex conditional responses depending on circumstances.

Neuronal interactions at synapses can be subtle because, over the short distances involved, neuronal membranes show "graded" responses. The neurotransmitter at an excitatory synapse causes a degree of depolariza-

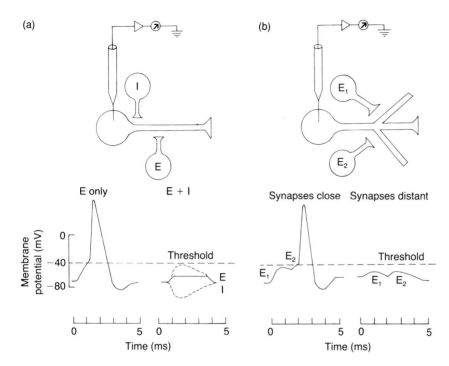

Figure 3.4
Interaction of presynaptic neurons to influence post-synaptic neuron.
(a) Excitatory and inhibitory synapses. If the presynaptic neuron forming an excitatory (E) synapse fires, the postsynaptic neuron also fires; but if both excitatory and inhibitory (I) pre-synaptic neurons fire, the postsynaptic neuron is inactive.
(b) Two excitatory synapses (E_1 and E_2). One alone may be insufficient to make the cell fire, but the two together may be sufficient. The exact interaction can depend on a number of factors, including the distance from the synapse to the firing part of the neuron. If the distance is short (left), the two excitatory synapses are more likely to stimulate the cell than if the distance is long (right). *(Modified after Young)*

tion of the postsynaptic neuronal membrane, but the membrane just after the synapse differs from the rest of the neuron and lacks the special properties that cause the controlled set-level depolarization. The post-synaptic depolarization has to diffuse passively down the cell until it reaches the part with the special neuronal membrane and, if the depolar-ization is large enough when it reaches that point, it will make the neuron fire into action. But the postsynaptic depolarization may be too small to make the neuron fire. Then if a second presynaptic neuron (which also has an excitatory relation with the postsynaptic neuron) fires close enough in time, the small depolarization it gives rise to in the postsynaptic cell will combine with the other small, graded depolarization and together they may be enough to make the cell fire. The same mechanism explains how the rate of firing of one neuron conveys information through the nervous system. If a second depolarization rapidly follows the first, the graded postsynaptic responses to the two are more likely to combine to make the postsynaptic neuron fire; if the two depolarizations are farther apart, the postsynaptic response to the first may have decayed away before the second arrives, and the neuron will be less likely to become active. Thus, at excitatory synapses, the more rapid the firing of the presynaptic neuron, the more likely it is the postsynaptic cell will respond.

There are further dimensions to synaptic interactions. The physical size of the gap at the synapse may vary, such that if the space is smaller the two neurons have a stronger influence on each other. One way in which learning may occur at the neuronal level is by adjustments to the width of synapses. A further mechanism uses the distance, in the postsynaptic neuron, between the synaptic junction and the region of the cell that shows the full on/off depolarization response (Figure 3.4b). The small graded depolarization has to diffuse to this region before it can have an important effect, and if the distance is longer, the neuron is less likely to respond: the small depolarization will have decayed away before it reaches the special membrane. Alternatively, the properties of the cell membrane in the synapse may vary to make the graded postsynaptic depolarization more or less easily propagated. The neuron forms synapses with prior neurons through a branchlike network of "dendrites," and by having branches of different lengths (or rate of conduction) to different presynaptic neurons, it can tune how quickly it responds to each of them. This property should be generally influential in nervous system interactions, and we shall meet a particular instance when we consider sensory perception (though it is not known whether the sensory system employs exactly the mechanisms described here, or an analogous physical mechanism that produces the same result). For completeness, we should also note that, in addition to the chemical synapses we have considered, there are also electrical synapses, in which the depolarization of the presynaptic neuron directly leads to a full depolarization of the postsynaptic neuron. Electrical synapses are particularly important in neuronal control systems that require a rapid response.

3.1.5 The neuronal control of singing in crickets

Let us consider an example of a behavior pattern that can be controlled by the action of motor neurons and interneurons alone. It is, at the neuronal level, one of the best understood of behavior patterns: singing in male crickets. Cricket songs are familiar sounds, because they can be heard at night in the summer throughout much of the world. Males sing in order to attract females. The crickets produce the sound by scraping their wings together (Figure 3.5), and the muscles used for singing are the same as those used for flying; many of the same neuronal circuits are also used in flying and singing.

The movement of wings is controlled by a set of muscles in the thorax (the thorax is the middle section of the insect, between the head and the

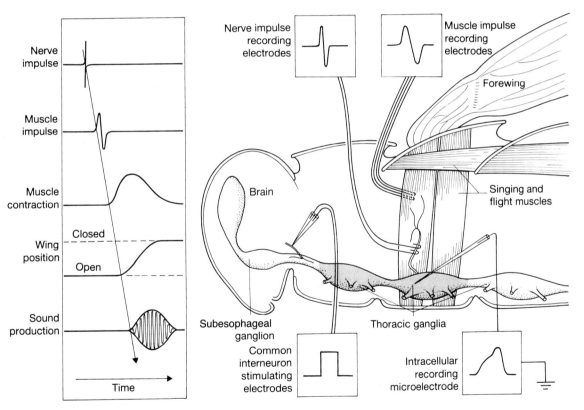

Figure 3.5   The nervous control of the song in male crickets. A normal song can be stimulated electrically in the command interneurons situated between the subesophageal and thoracic ganglia. The command interneurons control the interneurons of the thoracic ganglia, which in turn contol the motor neurons that control the muscles that effect singing, by means of scraping together the wings. *(From Bentley D, Hoy RR. The neurobiology of cricket song. Sci Am 1974; 231(2): 34–44. © 1974, Scientific American, Inc.)*

abdomen). Motor neurons run from the thoracic ganglion (a ganglion is a minibrain, an aggregation of many nerves) to the singing muscles. It takes many muscle fibers to move a wing, and each muscle fiber has its own motor neuron. All the muscle fibers must contract at the same time, and the coordination is produced by a "command interneuron," onto which all the motor neurons join. Soon after the command interneuron fires, all the motor neurons fire in unison. The coordinatory function of the command neuron has been proved by experiment. The command interneuron is situated in the cricket's neck. A neurophysiologist can locate it and join it to an electricity supply. If the electricity is switched on, all the motor neurons fire together, and the cricket produces a perfect song. It will produce its song even if it has had its head cut off. The command neuron, motor neurons, muscles, and wings form a complete singing machine.

The fact that a decapitated male cricket can sing normally, at least for a while, suggests that feedback from the external environment is not needed for it to produce its song. The song is not a response to an environmental stimulus (though singing is influenced by many environmental variables, such as temperature). The control of the song is by an internal rhythmic mechanism. The internal rhythm is probably generated in the thoracic ganglion, because when the command interneuron in the neck is electrically stimulated and causes the cricket to sing, the stimulus contains no timing information. The song consists of a regular series of sound pulses, delivered at a certain frequency in time; the stimulus from the interneuron in the neck probably serves to set the whole rhythmic machinery in motion but does not itself directly control the contraction of the wing muscles. The muscles contract rhythmically under the control of neuronal circuits in the thorax. Many kinds of behavior are controlled by internally generated rhythmic patterns rather than being produced in response to external stimuli: walking, swimming, flying, and behavior associated with daily (or longer) biological rhythms are all examples. However, other kinds of behavior are more directly stimulated by factors in the external environment. To understand them we need to move on to the third kind of neuron, which we so far have not considered: sensory neurons.

3.2   Sensory systems

3.2.1 Organs of sense

Animals need to be able to find their way around their environment, find food, recognize the species and sex of other individuals, even (in some cases) whether another individual is a member of the same or a different group, and detect the signals sent to them. For all these functions they rely on their sense organs. Different kinds of animals have different sets of sense organs. The set of sense organs possessed by each kind of animal is appropriate to the environment in which it lives. For example, species of fish and shrimp that live in dark, underground caves do not have eyes, or have eyes so reduced that they no longer work; there is no advantage in having light-sensitive organs where there is no light. The human set of eyes, ears, touch, and relatively poor taste and smell is just one particular, not very common, set of sense organs; most species of animals construct their perceptual worlds using sets of senses that differ from ours.

We can divide the different sense organs of animals into three groups: exteroceptors, enteroceptors, and proprioceptors. This division was suggested by Charles Sherrington at the beginning of the century. Exteroceptors sense the state of the environment outside the animal, enteroceptors the state of the animal's body, and proprioceptors the animal's movement by sensing the position of its muscles. Proprioceptors located in the cricket's wing muscles are important in the control of song, for example, as they sense the position of the wing. In mammals, enteroceptors include organs that sense the body temperature and chemoreceptors that sense the concentrations of chemicals such as hormones and carbon dioxide. The classic five human senses—sight, hearing, touch, taste, and smell—are effected by exteroceptors. But the division into exteroceptors, enteroceptors, and proprioceptors is not the only possible classification. Another possibility is to divide sensory systems into those that sense chemical, electromagnetic, and mechanical energy. Taste and smell are both chemical senses (as are many of the enteroceptors), hearing and touch are both mechanical senses, and vision is an electromagnetic sense.

There is no one "best" classification, and different classifications are useful for different purposes. Here we shall consider briefly some examples of electromagnetic, chemical, and mechanical sense organs. In the next two sections, we consider how sensory neurons work and look in detail at one example, echolocation in bats.

Let us first take a sense lacking in humans, the electrical sense. This has been most studied in fish. Their electrical sense is not the same as the sensitivity to pain by which we become aware of an electric shock, but another sense, comparable to our sense of hearing or smell. Some kinds of fish, such as dogfish (a member of the elasmobranch group that includes sharks and rays), use their electric sense to find food buried in the bottom sand by sensing the disturbance to the electric field that the buried living matter causes. The electric sense organs of elasmobranch fish are called the ampullae of Lorenzini and are a set of jelly-filled tubes beneath the skin. Various animals, including honeybees and some bacteria, can sense magnetism. Experiments we shall meet later (Chapter 5, section 5.3) suggest that at least some bird species can also sense the magnetic field.

The commonest electromagnetic sense is light vision. Light receptors, in the form of eyes, are found in more or less complex form in many kinds of animals. But eyes are not the only light-sensitive organs known in nature. Insects, for instance, have three light-sensitive ocelli on the top of their heads, behind their compound eyes. The functions of the ocelli are uncertain.

Chemical sense organs are found in many places in different kinds of animals. The sea hare *Aplysia* (Figure 3.6), which is a favorite animal for work on the nervous system, can smell seaweed, on which it lives. By recording the activity of the nervous system at different parts of the animal while it is smelling seaweed, it has been found that the chemical sense organs are in its tentacles. Houseflies have chemical receptors in their feet, which enable them to detect food (such as sugary water, in an experiment) by walking into it. Most insects have chemical receptors in their antennae. The antennae also contain mechanical sense organs, but an insect's whole surface has mechanical sense organs on it. Mechanical sense organs all work by means of tiny hairs. When the hair is bent a

Figure 3.6
A copulating pair of sea hares (*Aplysia*), off the Isle of Mull (in Scotland). They are hermaphrodites and fertilize each other. *(Photo: Dick Manuel)*

sensory neuron attached to it fires into action. Hearing works as a mechanical sense, and is also effected by small movement-sensitive hairs connected in some way (depending on the species) to a membrane that is set in oscillation by sound. Fish and some amphibians possess a special organ called the lateral line for detecting water pressure. The lateral line is a channel under the skin of each side of the animal, with little holes leading to the outside. The flow of water into the lateral line allows the fish to measure the movement of water with respect to itself.

3.2.2 Sensory neurons

Information is carried from an animal's sense organs to its central nervous system by sensory neurons (Figure 3.1). Sensory neurons are sensitive to certain kinds of stimuli in the external environment, and convert them into an electrical impulse—a wave of depolarization traveling from the sensory end of the neuron toward the brain. The conversion of the sensed form of the stimulus to electrical depolarizations carried in the nervous system is an instance of what physicists call "transduction." There are as many mechanisms of transduction as there are kinds of sense organ. Light-sensitive neurons in the eye, for example, contain chemical pigments that change in form when hit by light of certain wavelengths. It requires a special neuron to sense light. If you shone light on any other kind of neuron it would have no effect on it; only a neuron containing a light-sensitive pigment is set in action by light. The pigment's change in chemical form influences the property of another chemical that is wrapped around the pigment. The second chemical in turn influences a third, and so on, until we reach a chemical that influences the neuron's membrane, causing it to depolarize. Thus transduction is achieved by a chain of chemical reactions, beginning with a light-sensitive pigment and ending with chemicals that can depolarize the cell membrane. The chain of reactions is important because it acts to amplify the signal. The energy in a photon of light is small—too small to produce the electrical change needed to depolarize a neuron—but by triggering a prearranged cascade of chemical reactions, the photon can stimulate a light-sensitive sensory neuron into action.

3.2.3 The sonar system of bats

Bats are a large group of mammals that contains two subgroups. There are about 175 species of fruit-eating bats—the flying foxes and their relatives—and they (with one known exception) lack sonar and use eyesight to find their way around. The other subgroup, with about 800 species, contains the insect-eating bats; they are all thought to employ sonar. They have perhaps proliferated because their unique sensory skill allows them to exploit a resource (night-flying insects, particularly moths) for which they have almost no competitors. Bats can fly with ease in complete darkness; they do not collide with obstructions, and they catch their prey on the wing. They are capable of flying around a laboratory room that is crisscrossed with a network of wires of a diameter of as little as three hundredths of an inch, and catching flying moths from a range of eight feet. How do bats achieve this? The greatest experimental biologist of the eighteenth century, the Italian Lazzaro Spallanzani, could not solve the problem. He did find that if he stuffed up the bats' ears their ability to avoid obstructions declined, but he did not know how to explain this result. Bats remained a puzzle until after advanced equipment for recording and producing sound had been developed along with radar in World War II. The equipment was applied to the bat puzzle by Donald Griffin and his collaborators in the 1940s and 1950s. It was Griffin's group who established, first in the little brown bat (*Myotis lucifuqus*), that insectivorous bats find their way around by listening to the echoes of high-pitched sounds that they make themselves.

Humans cannot hear sounds much outside the frequency range 2000-20,000 Hz. (Frequencies of waves are measured in "Hertz" units, abbreviated to Hz. One Hz is one cycle per second. In the case of sounds, the higher the frequency, the higher the pitch.) Bats make sounds mainly of 20,000 Hz and higher—some emit sounds of up to 100,000 Hz—and most bat sounds are therefore inaudible to humans. One advantage to the bat of using such high-pitched sounds is that it is relatively undisturbed by background noise (most noises in nature have a frequency lower than 20,000 Hz). By concentrating on high-frequency sounds, bats live in a world silent except for their own noises and echoes. It is essential to the bat that there should be no interfering background noise. In an experiment, Griffin blasted noises of frequencies of more than 20,000 Hz into a room that contained various obstacles. Flying bats now bumped into the obstacles and fell to the floor. As well as showing the importance of a silent background, this experiment also gives part of the evidence that bats rely on a sonar sensory system.

Bats are not the only kind of animal to use sonar. Dolphins and other "toothed" whales, as well as the small mammals called shrews and a bird called the Malayan cave swiflet, which lives in dark caves, all echolocate. Also, echolocation is not the only function of the sonar apparatus. Dolphins and toothed whales are known to be able to stun prey by blasting them with intense bursts of emitted sound. However, the sonar systems of bats, and their use in echolocation, have been the most studied, and we shall concentrate on them here.

The sonar sound pulses of different bat species vary greatly, but we can distinguish two main kinds (Figure 3.7). Every species uses one or the other kind, or some mixture of them. One kind is frequency-modulated (FM) sound, in which the bat emits short chirps, each with a downward swoop of frequencies; the other kind is a longer sound with constant frequency (CF).

Figure 3.7
Two kinds of bat sonar: (a) constant frequency (CF) pulse of the greater horseshoe bat *Rhinolophus ferrumequinum*. (b) frequency-modulated (FM) pulses of mouse-eared bat *Myotis myotis*. The modulation increases as the bat approaches its target: from left to right, the pulses were emitted at distances of 4 m, 36 cm, and 7 cm from the target. *(After Neuwiler, Bruns, and Schuller and Habersetzer & Vogler)*

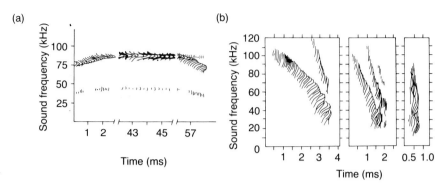

How do bats use these two kinds of sound? A long CF signal is useful for measuring the Doppler shift between the emitted sound and its echo. The Doppler shift is familiar to us as the sudden decrease in pitch we experience when a loud, rapidly moving vehicle such as an ambulance approaches, goes past, and then moves away from us. The extent of the shift is easiest to measure with a continuous tone of a single, pure frequency. When a bat is flying toward an object, emitting CF sound, the echo is Doppler shifted up in frequency. The greater horseshoe bat (*Rhinolophus ferrumequinum*) is a species that uses CF sonar (Figure 3.7a). It typically emits sound at about 80 Hz, and in this case the echo from a flying moth is at around 83 Hz (depending on the moth and the bat's exact relative motion). The bat could in theory use the amount of the Doppler shift to estimate the relative velocity of itself and its prey, in the same manner as the police estimate the speed of a motor car by radar, but there is no evidence that bats do this. Instead, bats are known to use CF sonar to detect the wing-beats of moths. As a moth moves its wings back and forth in the air, the Doppler shift in the sound echoed from them flutters up and down as the wing moves toward and away from the bat. This flutter in the Doppler shift is highly characteristic of moths, and makes them stand out against the background: the rustling of leaves and movement of twigs in the wind is much less regular than the moth's wing-beat. Thus bats can detect moths by listening (in a sense) to the fluttering of their wings.

Bats use FM chirps in another way. Each chirp is short, 2–3 msec or less (Figure 3.7b), and can be used to estimate the distance to a prey by the time delay between when the bat emits the sonar and when the echo returns (Figure 3.8). At an air temperature of 25° Celsius, for example, each one millisecond of time corresponds to about 17.3 cm of distance from the bat to the target from which the echo rebounds. Typically, as the bat homes in on its prey it increases the pulse rate and speed of frequency modulation in its sonar (Figure 3.7b), which is appropriate because the time interval of the echo shortens as the bat approaches its target.

The distance to the target is not the only information in the echo (Figure 3.8). The bat can also sense the target's size, direction, and height relative to itself. The target's size is proportional to the amplitude of the echo, because a larger target reflects more of the emitted sound, and is

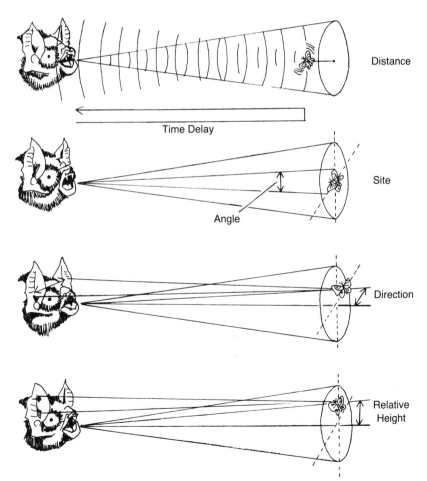

Distance

Time Delay

Site

Angle

Direction

Relative Height

Figure 3.8
Four kinds of information in bat sonar. The bat infers the distance to, and the size, direction, and relative elevation of, its target. These are respectively inferred from the time delay to the echo, the angle over which echoes are returned, the time difference between the echoes to two ears, and the interference between echoes to the top and bottom of the ears. *(Modified after Suga)*

also inferred from the angle subtended by the target in the sonar beam. Direction is indicated by the time delay in detecting the echo between the bat's two ears: if the moth is to the left, its left ear hears the echo a fraction of a second before the right ear. (Bats are extremely sensitive to short time intervals: experiments indicate they can resolve time intervals as short as 69–98 millionths of a second.) The height of the target relative to the bat is acoustically a more complicated inference. The bat has tall ears and echoes reflected to the top and bottom of the ear can "interfere" with each other; the pattern of interference differs with moths above, below, or at the same elevation as the bat.

Echolocation becomes ineffective beyond distances of about 30–40 meters, because sound is rapidly absorbed in air. But even for these short distances the bat must listen for a very faint echo of its much louder original pulse; the sound pulse emitted by the bat is about 2000 times louder than the echo. The bat therefore has the problem that it has to make a loud noise and then hear a soft noise immediately afterwards. When an ear has been blasted with a loud sound it becomes less sensitive for a second or so; human ears, for this reason, take some time to recover after listening to a very loud noise. Bats have two main methods of solving this problem. One is to make the ear less sensitive when emitting the sound. A nerve going to the muscle of the ear is automatically activated whenever the bat releases its sound pulse. This nerve causes the bat's ear to relax about 5 thousandths of a second before the pulse is emitted, and to recover about 10 thousandths of a second later. The ear will then be at peak sensitivity for picking up the echo. The other method is found in bats, such as the horseshoe bat, that use the Doppler shift. We saw earlier that the echo has a different frequency (maybe 83 Hz) from the original emitted pulse (at 80 Hz). The ear of the horseshoe bat is relatively insensitive to 80 Hz sounds, by a physical property of the tympanum, but is very sensitive at 83 Hz. The horseshoe bat is, so to speak, rather deaf to itself and sensitive to its own echoes.

3.2.4 The neuronal mechanisms of bat sonar

The research of Nobuo Suga and his collaborators has made progress in understanding the nervous mechanisms by which a bat infers the position of its prey. They have particularly studied the mustached bat *Pteronotus parnellii*. The bat hears the echo in the same way as all mammals detect sound. The sound waves strike the ear membrane and set the ear bones in motion; the movement of the ear bones is sensed by small hairs that lie in contact with them. The mechanical stimulation of the hairs causes a depolarization in sensory neurons contained in the auditory nerve. The auditory nerve runs to the brain, where the sense data is interpreted, particularly in a region called the auditory cortex.

Suga has investigated various components of sonar; let us concentrate on how bats measure the echo's time delay, a measurement that is crucial in estimating the distance to the prey. Microelectrodes in the bat's auditory cortex were used to record the response to pairs of pulses, separated by various time intervals, that were played to the bat. The pair of pulses corresponds to the bat's own chirp and its echo, but both were made artificially in the experiments. Suga and his colleagues found a series of neurons that were activated by pairs of sound pulses separated by distinct intervals. The neurons were "higher order" interneurons, several stages back from the sensory neurons in the ear; these interneurons construct the perception from the raw sounds. The neurons were not sensitive to single sound pulses—they require a pair—and an array of neurons was sensitive to various time delays (Figure 3.9): one set of cells

Figure 3.9 Perception of echo delay times in bat auditory cortex. (a) The auditory cortex is a region of the cerebrum in the bat brain. Suga and his co-worker have found a series of neurons in the auditory cortex that are sensitive to various time delays. (b) The neurons are arranged in a characteristic spatial pattern across the cortex; the positions of neurons sensitive to delays of 1–18 msec are illustrated. (c) The responses of six particular neurons to echo delays from 1–15 msec and various sound pressures. Each neuron responds to pairs of sound pulses inside the area shown, but not outside. *(Modified after O'Neill & Suga)*

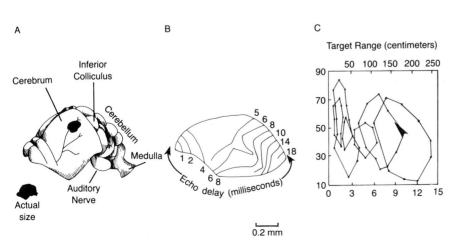

fired if the pair of sounds were 2 msec apart, another if they were 10 msec apart, etc. It is easy to imagine that, when a four msec echo-time neuron fires, the bat brain constructs the sensation "target at (about) two meters," and when a two msec neuron fires, it thinks "target at one meter," and so on.

Other experiments suggest how the lower neuronal circuits feed the sound information to those interneurons in the auditory cortex that we have just been considering. In natural conditions, the first sound pulse of the pair is made by the bat itself and is conducted through its head; the second pulse comes from outside, as an echo. It seems that the auditory cortex interneurons are not actually sensitive to time delay, but to synchronous stimulation by two prior neurons. The system may work as follows (Figure 3.10). The neurons that sense the first pulse, emitted by the bat, transmit the information toward the auditory cortex slower than do the neurons that sense the echo. There is a whole array of neurons that slow down the first (emitted) pulse by ½ msec, 1 msec, 1 ½ msec, etc., relative to the second (echo) pulse. If the actual time delay between the first and second pulse is 1½ msec, then the interneuron that is activated in the auditory cortex will be the one that responds when stimulated synchronously by two particular prior neurons: one at the end of a neuronal chain that sensed the first pulse and delayed it by 1½ msec and another neuron that sensed the echo. Likewise, the interneuron in the auditory cortex that responds to a pair of sound pulses 1 msec apart fires when stimulated by a neuron recording the echo and another that is at the end of a chain delaying the original pulse by 1 msec relative to the echo, and so on, for the other intervals.

We saw in the previous section some mechanisms by which neuronal information can be transmitted at various rates, according to the properties of the nerve cells and their synapses, and these mechanisms may operate in some form in the bat. We also saw how a neuron may fire only if stimulated by two prior neurons. The use, in sensory perception, of comparisons between pairs of stimuli, one delayed relative to the other such that the pair is sensed by synchronous signals to a higher order interneuron, is reasonably well-confirmed in the sonar system of the mustached bat. The mechanism is probably used in many other bat species, but its importance is broader still. There is evidence that analogous systems are used in human hearing, and the work of Hubel and Wiesel suggests there are similar mechanisms in the visual systems of cats and monkeys. It may be a general principle of sensory neurophysiology.

When we first learn about the sonar system of bats, it is tempting to imagine it, in human terms, as something rather crude: perhaps rather like going around in the dark, finding our way by clapping and listening

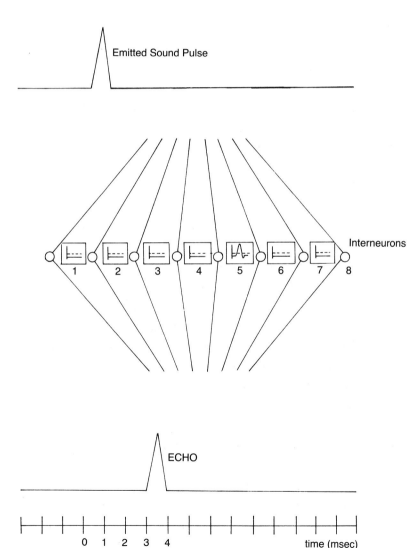

Figure 3.10

An abstract and hypothetical diagram of the nerves in the bat's perceptual system. The emitted sound pulse is sensed by one set of interneurons (upper half of picture) and the echo by another set (lower half). The neurons in each set delay the sound by different amounts, and a set of interneurons (1–8) can become active when simultaneously stimulated by the upper and lower neurons. Interneurons 1–8 are therefore sensitive to an array of echo delays. The circuits leading to interneuron 5 delay the emitted sound by 3–4 msec relative to the delay, and this interneuron is stimulated by the illustrated combination of emitted sound and echo.

for the echo. This would be a sensorily deprived world—like living in pitch black, the darkness punctuated only by sounds and echoes. It would be very different from the rich and beautiful sensory world provided for us in light by our visual system. However (as Richard Dawkins has argued), this conception may well be naive. The world we "see" is all constructed in our brains; it is constructed from the firing pattern of sensory neurons leading into the brain's visual cortex. The patterns we see, and the colors and textures, are all stimulated in our brains. The auditory cortex of bats is at least as big, relative to their brains, as the visual cortex of related mammals, and there is no reason why bats should not be able to make an equally sophisticated perceptual model of the world. The pattern of echoes reveals the shape, the position, the movement, even the wing flutters of moths, and the rustling of leaves and swaying of branches in the surrounding trees. Maybe the bat's brain treats auditory information in the way we treat visual information, using the information on movements and shapes to construct a mental image of the world. A bat could even paint onto its mental image of the world (as we do) colors corresponding to the different kinds of objects it distinguishes. The bat could then have as rich a mental perceptual world as we do, but reconstructed from a different sensory source.

The neurophysiological work has revealed that the bat's auditory cortex does operate in the way that, on general grounds, can be expected of a system that is constructing an image of the world. A key requirement in constructing an image of a scene is that events in different places must be represented simultaneously. The echoes from distant objects actually arrive after the echoes from nearer objects, and the simulation of an image would require the information from the nearer objects to be slightly slowed down. Maybe some extension of the abstract model of Figure 3.10, which is about detecting a single echo, is at work. Maybe different echoes from objects at various distances could be put through an analogous array of still higher order interneurons such that they could be simultaneously perceived deeper in the brain. In 1993, S.P. Dear and colleagues described a set of neurons in the auditory cortex of the big brown bat (*Eptesicus fuscus*, Figure 3.11) that behaved in the right way: they transformed the sequentially arriving echoes from different objects, at different

Figure 3.11
A big brown bat
(*Eptesicus fuscus*),
catching a suspended
mealworm.
(Photo: Dr. Steven Dear)

distances, such that they could be simultaneously represented. Thus the
bat brain functions as if it were building up an image of the world.

Nevertheless, we do not know how bats subjectively experience their
external world. The main value of Dawkins's argument is to point out
that what we perceive as an objective external world is actually a simulation, constructed inside our heads from an electrical firing pattern of
neurons. We have no reason to generalize from the way we use visual
and auditory information to other species; we happen to construct a rich
image from visual information but not from auditory information. But

bats have about as rich a set of information in their auditory interneurons as we do in our visual interneurons, and therefore it is possible they reconstruct the external world in as much detail as we do.

3.3   Sensory information and behavioral decisions

Animals are selective in the environmental information they respond to. The environment of any species contains far more information than an individual could ever use in deciding on a course of action. We can think of the selection as acting at three levels. At one level, sense organs simply do not pick up the signal. For example, the external environment contains a whole spectrum of electromagnetic rays; in addition to the small band of frequencies we perceive as colors there are x-rays and radio waves, together with other high- and low-frequency radiation of which we are not normally aware.

At a neuronal level, there is the interesting fact that some parts of the light patterns that do stimulate the retina play no part in forming the perception in our brain. The reason is that it is inefficient to concentrate on all the available sensory input when forming an image of it. An object in the environment (such as a tree) will stimulate a whole column-shaped region of the retina. But only the information about the edges of the object needs to be sent to the brain: the sensory neurons corresponding to the middle of the object are ignored. The whole shape of the object can then be reconstructed in the visual cortex of the brain; it just needs to know where the edges are, and the rest can be filled in without information from the sense organ. If you imagine the brain scanning across the retina, it would clearly be more economical for the retina to send a message saying "retinal region 2: tree begins here—retinal region 14: tree ends here" rather than one saying "retinal region one: no tree—retinal region two: tree present—retinal region three: tree present—retinal region four: tree present—" with information for all the regions being sent. At a neuronal level in the eye, the process responsible is called "lateral inhibition." If there are three neurons next to each other and the two outside ones are active and depolarized by a stimulus, then their action inhibits the middle one. Then even if the same light pattern is hitting the middle neuron as the outside ones, only the outside ones send a signal to the brain. Lateral inhibition is a widespread property in sensory neurophysiology; eyes are not the only place where it operates.

Thirdly, there is the principle called "stimulus filtering," by which the sensory system of an animal responds to some stimuli in the environment but not to others—or to some aspects of an object rather than others. How can we find out what stimuli an animal is responding to? One method is neurophysiology. We can record the activity of sensory neurons when an animal is successively presented with a variety of objects. We shall meet one example of this kind of study in the prey-catching behavior of the toad (p. 143). The same kind of question can also be tackled by a higher level method. We can ignore the physiological intermediary details, and find out what kind of environmental stimuli the animal responds to at the behavioral level. The simplest kind of sensory information is used in the responses called "kinesis" and "taxis." In a kinetic response, the animal alters its rate of movement, in a random direction, according to the intensity of the stimulus. When the stimulus, which might be light or moisture, is of the right intensity it slows down and thus spends more of its time under those conditions. Woodlice show a kinetic response to moisture. They move faster where it is drier, and therefore spend more time where it is moist. The unicellular organism *Paramecium* shows a kinetic response with respect to the local concentration of carbon dioxide (Figure 3.12). A taxic response, however, is directional. Negative phototaxis, for example, means that the animal moves away from light, as in fact does the maggot of the bluebottle fly. The taxic response is made by sensing the direction of the light, which is achieved by different techniques in different species. It requires only the most elementary kind of sensory information. The animal does not need to know anything about the light source, only that it is light and where it is coming from.

Few behavior patterns, except perhaps chemical responses to food (or pheromones, see p. 169), are guided simply by the intensity of one stimulus. More of the behavior patterns that we can see animals performing are controlled also by the pattern of the stimulus in the environment. For example, in the case of a visual stimulus, it is not only the presence or absence of light that matters, but the exact distribution of degrees of

Figure 3.12
Response of the uni-cellular organism *Paramecium* in the vicinity of a CO_2 bubble. When it senses a high CO_2 concentration it withdraws, turns through a certain angle, and advances again, but its new direction is unrelated to the direction of the stimulus. *(After Kuhn)*

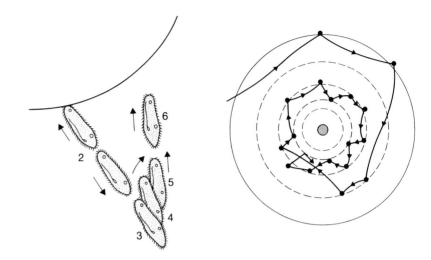

light and shade, which define the shape and appearance of the image. Let us consider the egg-retrieval response again, this time as an example of a more complicated analysis of the exact stimuli that influence behavior. We saw in Chapter 1 that a goose will respond to an egg outside its nest by rolling it in. Gulls do likewise. But what exactly is it that stimulates the behavior? How does a gull recognize an egg? Gerard Baerends and his colleagues sought to answer this question by experiments with model eggs of various size, stippling, shape, and color. They presented herring gulls, on their nests, with choices of different model eggs, and recorded which ones were preferentially retrieved (Figure 3.13). The gulls preferentially retrieve larger models, even if their size exceeds the natural size of an egg; the size of an egg can therefore be used as a scale of comparison for the other variables. Baerends tested each model egg by giving gulls a choice between the model egg (which might be varied in its color, shape, or stippling) and a range of sizes of control models (of normal shape, color, and stippling). Figure 3.14 summarizes the results. The gulls take more notice of some variables than of others. They take little notice of shape: an oblong model is as likely to be retrieved as an egg-shaped model of the same size (compare *a* with *R* in Figure 3.14). Nor do the gulls prefer eggs of natural coloration: they preferentially retrieve a green egg rather than a naturally colored egg of the same size (compare *c*

Figure 3.13
A black-headed gull with a choice of an artificially large egg (of natural color and stippling) and a natural egg. The gull is retrieving the large egg. *(Photo: Niko Tinbergen)*

with R); but they are sensitive to color, as they prefer green to brown eggs (compare b with c). The other variable, besides size, which strongly influences the gulls' preference is stippling: they prefer stippled to unstippled eggs (compare d with c). It is as if natural color and shape are not part of a gull's idea of an egg, but size and stippling are. They do not simply prefer to retrieve eggs of natural appearance, and indeed prefer larger than normal eggs, in which case the models take on the condition of a "supernormal stimulus": the model is a more attractive stimulus as the stimulatory property is exaggerated. The supernormal stimulus in a sense deceives the egg recognition mechanism of the gull. In nature, the whole sensory mechanism works to distinguish eggs from other objects, but it can be tricked by experiment. Baerends's experiment is a neat example of the analysis of a stimulus; its main conclusion is that herring gulls recognize eggs mainly by the criteria of size and stippling.

"Behavioral assays," such as the egg-retrieval response of birds, not only reveal what stimulus pattern is recognized by the animal; they are also a revelatory method of studying the sensory powers of animals. If an animal can be shown, by appropriately controlled experiments, to behave in response to some property of the environment, it must be able to sense it. Karl von Frisch applied the method to demonstrate the hearing ability and color sensitivity of fish. In that case, the physiologist von Hess had asserted that fish are colorblind and deaf; von Frisch doubted

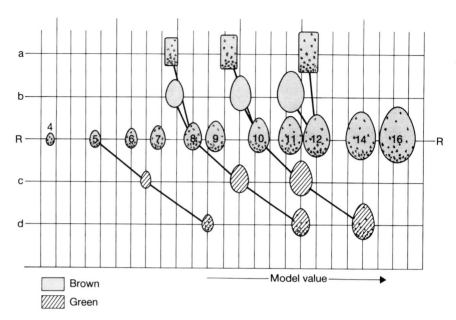

Figure 3.14   Preferences of herring gulls for different model eggs. Herring gulls were offered, for retrieval, a choice between a model egg of a certain size, shape, color, and stippling, and another egg of natural appearance but not necessarily of natural size. Gulls prefer to retrieve larger eggs, and a scale of increasing egg size can be used as a scale to measure how the gulls assess other properties of the egg. The scale of increasing-sized eggs is the R-series in the middle of the figure; the number in the egg indicates its size relative to natural size in units of ⅛ (8 is natural size, 16 twice natural). There are four experimental series: (a) brown, stippled, block-shaped; (b) brown, unstippled, egg-shaped; (c) green, unstippled, egg-shaped; (d) green, stippled, egg-shaped. The preference for a model egg can be read off the graph by comparing its position on the horizontal axis with an egg of equal size in the R-series. For instance, a-series eggs (block-shaped) are displaced a little to the left of the R-series, which suggests that herring gulls take little notice of the shape of the eggs. Eggs of equal size are connected by thick lines for ease of comparison. *(After Baerends and Kruijt)*

the assertion, and he successfully trained minnows to distinguish colors by rewarding them with food, and catfish to come out of a tube when he blew a whistle. In both experiments he used a behavioral response to discover a sensory ability.

In this section we have considered behavior patterns that are elicited by one stimulus. However, in many cases a number of stimuli interact, or a stimulus may influence one behavior but not another. Niko Tinbergen, in classic experiments on a species of butterfly called the grayling (*Eumenis semele*), showed that the butterflies responded to flower color, but the response depended on the presence of scent (the scent of flowers). A butterfly's response to color also depends on whether it is behaviorally concerned with feeding or reproduction. Grayling males do not respond to color when seeking females, and therefore they must in some form surpress their sense of color in a reproductive context. In general, the behavioral decisions of animals depend on a variety of external sensory factors and internal motivational factors; the various influences are integrated together by an animal's central nervous system to control its behavioral output. The full problem of decision making is complex, but we shall consider some principles in the next section.

Before we move on, it is worth noting that not all behavior is controlled by the central nervous system (though most of it is). The main exceptions are "peripheral reflexes" (of which the human knee jerk is an example), which are controlled by a simple system of one sensory neuron and one motor neuron. The sensory neuron connects directly to the motor neuron, which in turn controls the muscles that effect the behavior pattern. When the sensory neuron fires (after mechanical stimulation in the case of the knee jerk reflex), it stimulates the motor neuron, which causes certain muscles to contract. We shall meet some other examples of reflexes, but in most behavior patterns the sensory input and behavioral output of the animal are less directly connected, and the senses exert their influence on behavior through the central nervous system.

3.4　Choices among behavior patterns

The behavioral output of an animal emerges as a sequence of many different behavior patterns, and each change in the sequence can be thought of as a behavioral "choice" made by the animal. The question is how animals make those choices. Some will be responses to changed environmental stimuli. New sensory information is one factor causing an animal to choose one behavior pattern rather than another, but it is not the only one.

The same animal may not respond to the same stimulus in the same way on different occasions, and it may change what it is doing even while the environment appears to be constant. There are two related reasons why an animal may behave differently when under similar environmental conditions. One is that its internal tendency to behave in a certain way may change. For example, when food is presented, it will become less likely to feed as it grows less hungry. The other reason is the interaction of behavioral preferences; an animal may stop feeding in order to avoid a predator. The two reasons are related because the internal tendency of an animal to behave in a certain way is presumably determined by a balancing act among the consequences of all its possible responses to a given environment. The balancing act, and the resultant behavioral preferences or tendencies, are called the "motivation" of the animal. Let us consider how the motivation of an animal influences its behavioral choices.

A study by Baerends on the freshwater fish called guppies (*Poecilia reticulata*) provides a clear example of the interaction of motivation and external stimulus. The behavior in question is the courtship of females by males. Baerends recognized three different behavior patterns that a male may perform when courting a female: "posturing" in front of a female, a limited sigmoid movement, and a full sigmoid display. How does a male decide which to perform? The answer seems to depend on the size of the female and the male's own motivation to court, which can be independently measured by his coloration. Figure 3.15 depicts the combinations of these two factors necessary for a male to court a female in each of the three ways. The particular shape of the graphs is not important here; they are only to illustrate a general point, which is that, for any behavior pattern in any species, there will be some such graph of motivational tendency and external stimulus that describes the conditions under which it is performed.

The guppy illustrates choice among different behavior patterns of one class, courtship. What of interactions among different kinds of behavioral goals? Here we need a new example, which (unlike the courtship of guppies) is understood neurophysiologically. The study of the gastropod *Pleurobranchia* by J.W. Davis and his colleagues has partly uncovered the neurophysiological control of six behavior patterns: feeding, egg

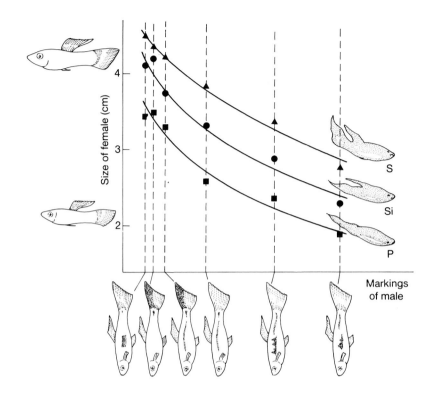

Figure 3.15
The tendency of male guppies to court females is determined by an internal factor—how inclined the male is to court (which is indicated by his color markings), and an external factor—the size of the female. The curves connect points where males will perform each of three kinds of courtship display, in order of increasing intensity of courtship: posturing (P), sigmoid intention movements (Si), and full sigmoid displays (S). *(After Baerends et al.)*

laying, escape, withdrawal of the oral veil, righting, and mating (Figure 3.16). Take first the interaction of feeding and egg laying. *Pleurobranchia* is a carnivorous snail that includes eggs in its diet. When a *Pleurobranchia* lays its own eggs, it switches off its feeding habit. Another behavior pattern, escaping from predators, is performed in preference to all other activities. A fourth behavior pattern is to withdraw its oral veil on being touched. A snail with its oral veil withdrawn cannot feed, and its tendency to withdraw its veil interacts with its tendency to feed. If food is abundant, or the snail is not hungry, withdrawal has priority over feeding, and vice versa when food is scarce and the snail is hungry. The other two activities studied by Davis are "righting" (turning the right way up) and mating. Having established the behavioral priorities by observation, he proceeded to their neurophysiology. The system has not been completely elucidated, but it is now known for instance that two neurons are responsible for inhibiting the "withdrawal" response when a

Figure 3.16

The behavioral priorities of the intertidal snail *Pleurobranchia*. "Escape," for instance, is performed in preference to "egg laying" when the circumstances appropriate to both behavior patterns are fulfilled. *(After Kovac and Davis)*

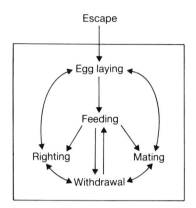

A → B   A dominates B

A ↔ B   A and B can be simultaneous

A ⇄ B   Reciprocal inhibition between A and B

Pleurobranchia is feeding, and that the inhibition of feeding during egg laying is effected hormonally. The priorities of *Pleurobranchia*, by the way, do make sense; without them it would eat its own eggs after laying them, and if escaping from danger did not have absolute priority, it would not survive to exercise its other behavioral preferences.

3.5   Hormones

The nervous system controls the behavior of animals over a short time scale, of matters of seconds or microseconds. Other factors control it over longer terms of days, months, or even years; the most important of these are hormones. Some hormones act quickly, but we shall concentrate here on those with a relatively slow effect. Hormones are chemicals that circulate in the bloodstream of animals, regulating metabolism and behavior. They are released by special glands, such as the pituitary gland at the base of the brain, and the gonads (the ovaries in the female, the testes in the male). The glands release their hormones into the blood; some other organ will then respond to the increased level of hormone circulating in the blood. The responsive (or target) organ is often, but not always, the nervous system. An example of a target organ other than the nervous system is the effect of the hormone testosterone in the African clawed toad

(*Xenopus laevis*). Testosterone is released by the male clawed toad's testes. An increase in the amount of testosterone in the clawed toad's blood stimulates the growth of special "nuptial pads" on the male's front legs, which the male uses to embrace the female while mating.

We can illustrate the operation of hormones by a classic study, of the reproductive cycle of the Barbary dove (Figure 3.17). It was worked out by Daniel Lehrman and his colleagues. The Barbary dove's reproductive cycle lasts about six to seven weeks. Before the beginning of the reproductive season, the level of testosterone in the male's blood is low. Male Barbary doves with low levels of testosterone are aggressive to females. The aggression of the male in turn suppresses the release of reproductive hormones in the female, which ensures that she does not become ready to reproduce before the male. The reproductive season comes on as day length increases; its beginning is determined hormonally, as the increase in the number of hours of daylight stimulates the male's testes to release testosterone. The testosterone acts in the Barbary dove's brain. In the male, it causes him to cease being aggressive to the female, and to start courting, which consists of a ceremony of bows and coos. That testosterone is responsible for the change can be shown experimentally by injecting it into a male who will soon start courting, if given a female. Males injected with another hormone, estrogen, show the same response, because testosterone is converted into estrogen in the brain before it exerts its effect of inducing courtship. Courtship is therefore stimulated by testosterone released from the male's testes, but after it has been converted into estrogen on the way.

The bowing and cooing of the male Barbary dove influences the hormonal system of the female. It stimulates her pituitary gland to release a hormone called gonadotropin, which stimulates her ovaries to grow, in preparation for egg production. The stimulated ovaries in return release estrogen, which stimulates the female to start building a nest. After a few days the nest is built and the female then lays two eggs. Meanwhile, the nest building will have stimulated, in both male and female, the release of the hormone progesterone, which brings on the next stage of the reproductive cycle—sitting on the eggs. This stimulates the release of yet more progesterone, and of another hormone called prolactin. Prolactin

(1) Aggressive pre-reproductive phase

(2) Courtship

(3) Nest building

(4) Incubation

(5) Parental care

Figure 3.17   The sequence of reproductive behavior of the Barbary dove is controlled by, among other things, hormonal changes. The initial response of a male Barbary dove to a female is to attack her (1), but as his testosterone level rises, he comes to court females (2). The male and female then cooperate to build a nest (3), incubate the eggs (4), and feed the young (5).

causes the changes in the male and female needed to make them able to feed their young. Barbary doves feed their young on a special kind of "milk." Dove "milk" is made in the crop (a region of the gut between the mouth and the stomach); bits of the lining of the crop break off, and are later regurgitated to the young as "milk."

The behavioral changes through the reproductive cycle of the Barbary dove are thus controlled by a series of five hormones: testosterone, estrogen, gonadotropin, progesterone, and prolactin. Each hormone stimulates the next stage of behavior, which in turn stimulates the release of the next hormone. But the release of hormones is not only controlled by the behavior of the animal releasing them; the hormonal system is also influenced by the external environment (as in the effect of daylight on the male) and the behavior of other animals (as in the effect of male aggression on the female). We shall meet some further examples of hormonal influences on behavior in later chapters (p. 223).

3.6   Summary

1. Behavior is controlled by the nervous system, which is made up of cells called neurons and operates electrically and chemically. There are three main kinds of neurons: motor neurons, which attach to and control the contraction of muscles; interneurons, which connect among other neurons in the body's neural network; and sensory neurons, which are sensitive to stimuli in the external environment.

2. Neurons are connected with each other at structures called synapses, and it is the various kinds of relations between neurons at synapses that enable the control of flexible and complicated behavior.

3. The song of male crickets is produced by the muscles that move the wings, and the synchronous contraction of all the fibers in the muscles is coordinated by a "command interneuron" in the thoracic ganglion.

4. Bats have a sonar system that enables them to locate objects in the environment by the echoes of sound pulses emitted by the bat. The emission frequencies and kinds of sound pulses are appropriate for various kinds of echolocation. Some of the neural mechanisms have been investigated: for example, the interval between the pulse and the echo is sensed by neurons in the auditory cortex, and the mechanism by which the cells sense the time interval is partly understood.

5. Stimulus filtering is studied by experimentally altering the properties of an object, such as a bird's egg, to see which factors the animal is responding to.

6. Behavioral choices are controlled by the nervous system: it integrates external stimuli and internal tendencies, and some behavior patterns have priority over others. The neuronal control of behavioral choices is partly understood for choices among six behavior patterns in the snail *Pleurobranchia*.

7. Hormones are chemicals that are released by certain organs and that influence other organs elsewhere in the body. Hormones can increase, or decrease, the chance that particular behavior patterns will be performed. The release of a series of five hormones controls the sequence of behavior in the reproductive cycle of the Barbary dove.

3.7   Further reading

Young (1989) explains the neural control of several kinds of behavior, including two of the main examples (cricket song and bat echolocation) in this chapter. On cricket song, further articles are by Bentley and Hoy (1974) and Huber and Thorson (1985), and Dethier (1992) describes the natural history, particularly for the Northeast. On bat echolocation, see Suga (1990). (See Au (1993) on dolphin sonar.) R. Dawkins (1986) is the reference for the possible perceptual world of bats, and Dear, Simmons, and Fritz (1993) is the neurophysiological paper mentioned in the text. The question of subjective experience in animals is an interesting topic to follow up, though not covered in this chapter: Griffin (1992) and M.S. Dawkins (1993) are two stimulating and readable books about it. The early work on the control of behavior, and on the use of models to study pattern recognition, is still as easily read about in Tinbergen's classic *The Study of Instinct* (1951) as anywhere. Manning and Dawkins (1992, Chapters 3 and 4) also discuss behavioral work on stimuli and decision-making. Colgan (1989) introduces the science of motivation, and McCleery (1983) discusses Davis's work on *Pleurobranchia* in the context of an introduction to behavioral decision making. Several chapters in section III of Bateson (ed., 1991) are on the theme of this chapter. Becker, Breedlove, and Crews (1992) contains recent reviews on behavioral endocrinology; Lehrman (1964) explains his work on the hormonal control of reproductive behavior in Barbary doves.

CHAPTER 4

The Genetics and
Development of
Behavior

Behavior patterns develop by the interaction of inherited pre-dispositions and experience. We begin by seeing how to study the genetic control of behavior, and then move on to the influence of experience in various forms of learning. We end by looking at a conceptual controversy concerning the term "instinct," and at the particular kind of learning that enables cultural transmission in nonhuman animals.

4.1   The principles of genetics

Inheritance has been the subject of some of the most important scientific discoveries of the past century. The mechanism of inheritance was cracked using nonbehavioral traits, but we can reasonably infer that behavior is inherited in much the same way as are other properties of organisms. It will be easiest to explain the principles of heredity with the traits in which they were originally discovered, and only then to turn to behavior. Behavior has been less studied than some parts of morphology, and its inheritance is often too complex to lay bare the elementary principles.

Genetics is now an advanced science, but we only need consider its elements. The first important discovery was made by Mendel in the mid-nineteenth century. He experimented on pea plants. He first isolated pure lines of eight paired traits, of which we shall confine ourselves to the lines of "tall" and "short" peas. (A pure line is one that breeds true for that character; when two peas from a pure "tall" line are crossed, they always produce a tall pea.) Mendel crossed tall peas with short peas and found that the first generation peas were always tall. He then crossed together members of the first generation, and found that in the second generation tall and short peas occurred in the ratio 3 tall:1 short. The explanation of Mendel's result is as follows (Figure 4.1). There are two factors, which we call genes. One causes tallness, and may be called *A;* the other causes shortness, and may be called *a*. Each pea has two such genes. Peas with *AA* are pure tall and peas with *aa* are short. (*AA* and *aa* were the two starting pure lines.) When peas are crossed each pea puts only one of its two genes into its offspring. Because there are two parents, the offspring receives two genes controlling size (one from each parent). All the first generation peas in Mendel's experiment received one *A* from their tall parent and one *a* from their short parent. They were, therefore, *Aa*. All these peas were tall. This is explained by *A* being "dominant" to *a:* if an organism contains two genes, one dominant to the other, then it develops the character controlled by the dominant gene. (The other, unexpressed gene is called recessive.) Now consider what happens when two of the first generation tall peas are crossed. They are both *Aa,* and their genes can combine in four different ways. An *A* from one parent can combine with an *A* (giving *AA*) or an *a* (giving *Aa*) from the second parent, and the *a* from the first can combine with an *A* (giving *Aa*) or an *a* (giving *aa*) from the second. *AA, Aa,* and *aa* are therefore produced in the ratios 1:2:1. However, *Aa* peas look like *AA* peas—they are both tall—so the ratio of tall:short is 3:1.

Inheritance, then, is effected by paired units, called genes, which determine the characters of the organism. The genes *A* and *a* that control size in peas are just one alternative pair among the many genes in a pea plant. Each organism contains thousands of genes; a human, for instance, possesses approximately 100,000 genes. Genes are the main hereditary material passed on from parent to offspring, but they do not completely determine what the offspring looks like. In technical language,

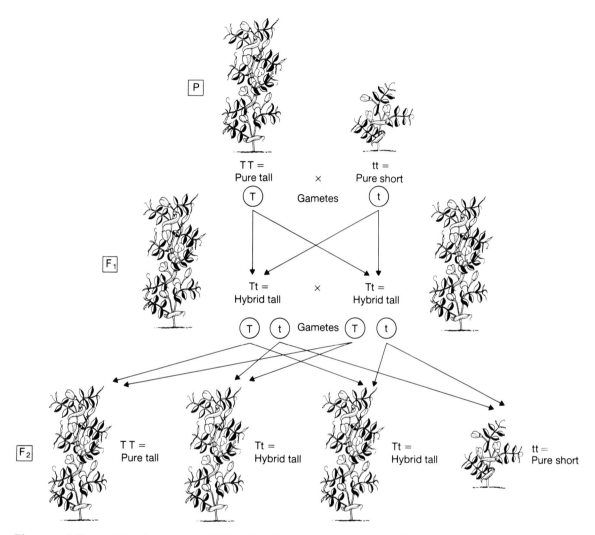

Figure 4.1   Mendel, in one of his original experiments on heredity, crossed a pure line of "tall" peas with a pure line of "short" peas. The first generation peas were all tall. He then crossed members of the first generation, and in the second generation the peas were tall and short in the ratio 3:1. He explained the result as follows. Each pea has two factors controlling height, but only one is passed on to its offspring. Each factor may be of one of two kinds, causing either tallness or shortness. The tall pure parental line had two tallness factors, the short line two shortness factors. Their offspring had one of each but were tall rather than intermediate in height because the tallness factor is "dominant" to the shortness factor. When the hybrid first generation peas were crossed, the tallness and shortness factors, in all possible combinations, produced ratios of three tall plants to each short one. The Mendelian factors are now called genes.

the difference between the genes that an organism inherits and the actual traits that it possesses, including its behavior, is the difference between its genotype and its phenotype. In the Mendelian example, the genotypes were *AA, Aa,* and *aa;* the phenotypes were "tall" and "short." We can therefore immediately see that there is not a 1:1 relation between the two: the class of tall individuals is made up of individuals with two genotypes, *AA* and *Aa.* In this case, the 1:1 relation has been broken down by dominance, but there are many other reasons why phenotypes and genotypes do not exactly correspond. The most important, as we shall see, is the influence of the environment on development.

The means by which genes produce their effects are known in some detail. Genes themselves are inherited in long chains called chromosomes, made of the chemical deoxyribonucleic acid (DNA). (Chromosomes can be seen with the aid of a microscope.) Genes directly dictate the manufacture of molecules called proteins, and it is through their proteins that genes exert their effects on the phenotype of the body. Although we know much about how genes work at the microscopic level, less is known about all the intermediate processes through which genes find expression in phenotypic characters at the macroscopic level of the whole animal. We possess no worked example to illustrate how genes control behavior; all we can say is that, in abstract terms, proteins influence the development of neurons, and that genes must by this means influence the development of the nervous system. That, however, is a vague assessment, and must remain as such, because the mechanism by which genes build bodies is one of the great unsolved problems of biology.

Size in peas, in the form Mendel experimented on it, was an exceptional trait; most traits, particularly behavioral ones, do not come in simple discrete pairs. Many traits, such as size in humans, vary continuously, or at least come in multiple forms (i.e., humans are not either short or tall, but a wide range of heights). Mendel in fact selected the paired categories of size because he reasoned, correctly, that the mechanisms of inheritance would be most easily revealed in such a trait. Continuously varying traits differ from size in Mendel's peas in that their variation is probably controlled by a large number of genes, and by environmental differences too.

Inheritance can be studied in continuously varying traits, but slightly different techniques are needed—the techniques of quantitative genetics. For a continuously varying trait (let us use human size as an example), the value of the trait in an individual is probably determined by a large number of genes, together with the effect of the environment. In the case of Mendel's peas size was controlled by two forms of one gene. The alternative forms of a gene are called "alleles." Each individual inherits two copies of each gene and therefore can have at most two different allelic versions of it. However, in the population as a whole there may be more than two versions of the gene, scattered among different individuals (e.g., one individual may be *Aa* and another *AA'* and the population would have three alleles of the gene, *a, A',* and *A*). In humans size may be controlled by multiple alleles of many dozens, or hundreds, of genes. (I should stress that I am using human size as an imaginary example: I do not know how many genes and alleles are really at work in this case.) The whole genotype controlling size in an individual would then be impossibly difficult to work out by Mendelian breeding experiments. Imagine for instance a case where size was influenced by 22 genes, each with only two alleles (*A/a; B/b; C/c . . . T/t; U/u*); and that each capital letter allele added three inches to the individual's height, whereas each lower case allele added nothing. An individual with two capital letter alleles at all 22 genes would then be $2 \times 3 \times 22$ inches = 11 feet tall. An individual (in this imaginary example) with lower case alleles of all 22 genes would have zero height. The genetic difficulty concerns all the intermediate heights. Take, for example, an individual who is 5½ feet tall. His genotype could be *AaBbCcDd . . . Uu,* but it could also be *AABBCC. . .JJKKLLmm. . .ttuu* or any other combination of genes such that half were capital forms and half lower case forms. These are a large number of possibilities and they would be practically impossible to distinguish by breeding experiments. In a Mendelian experiment on pea size we can find out what genotype an individual has, whether it is *AA, Aa,* or *aa.* For a trait influenced by more than a small number of genes we cannot work out the genotype by a Mendelian experiment. If we are to study the genetics of quantitative characters we should aim to know less than the exact genotype responsible for each phenotype.

In practice, although the exact genotype of an individual cannot be found out, we can find out something more abstract if less informative. We can find out the "heritability" of the trait. Heritability is a number between zero and one that expresses the extent to which the variation of the trait is due to variation in genotypes and in environments. In terms of size in humans, some differences between individuals will not be caused by differences in their genes at all, but by differences in how they grew up, how much they ate, and so on; other differences will be genetic. Thus, if we take the population as a whole we can ask what proportion of the variation in size is due to genetic variation and what proportion to environmental variation. Heritability is defined as the proportion due to genes; it is the ratio of genetic variation to the total variation. Examples of the two extremes may clarify the meaning. Heritability is zero if all the variation is nongenetic. For example, differences in the languages different humans speak are probably entirely caused by a nongenetic factor, in this case the language spoken by those around us when we are young. The heritability of human language is probably zero. At the other end of the scale, blood groups in humans are an example where heritability is one, because all the differences among individuals are due to their genes. Few traits reach the extreme of a heritability of one.

The important point about heritability is that we do not need to know anything about the actual genotypes in order to say what it is. The heritability of a character is independent of the genetic details. It is a more abstract, cruder kind of knowledge. We could calculate the heritability for a character of which we did know the exact genetics (such as Mendel's peas), but there would be little point in doing so, because the actual genotypes controlling the trait are more informative than the heritability. Measuring heritability comes into its own for those characters for which we do not know the genes involved.

So far I have only said what heritability is. How can it be measured? The fundamental technique is still the controlled breeding experiment. There are two important forms of breeding experiments. One is to breed together different parental lines that differ for the trait of interest. If we wish to know the heritability of size we breed together sets of parents of different sizes, and measure the size of the offspring. If there is a genetic influence, the offspring will, on average, resemble their parents. In practice we should draw a graph of the parental values against their offspring values, and if the graph shows any relationship the character is heritable. The strength of the relationship indicates the degree of heritability; if there is no relationship at all, heritability equals zero (Figure 4.2). (The method works with other classes of relatives as well as parents and

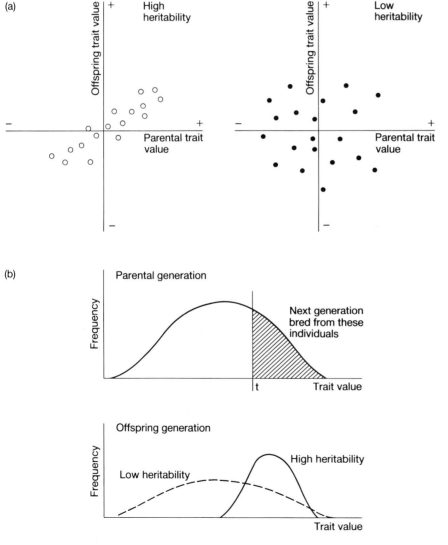

Figure 4.2

Measurement of heritability: (a) by breeding experiment. The value of the trait (which might be some quantitative variable such as size) is measured in the offspring and their parents; if the values are correlated, then (provided that the offspring have not been reared in a similar environment to their parents) the trait is heritable. (b) by artificial slection. Only those organisms of more than some value (*t*) of the trait are allowed to breed. If the average value of the trait in the offspring is higher than the parental generation average, the trait is heritable.

offspring. If a trait is heritable, all classes of relatives will be correlated to some extent.)

The second technique is the artificial breeding experiment. Artificial selection means breeding only from a selected minority of the members of a population. It is frequently used in agriculture. If we wish to increase growth rate in pigs, for example, we can try breeding only from the fastest growing pigs, and, if growth rate is heritable, a whole population of fast-growing pigs may be obtained in a few generations. If artificial selection does change the average condition of the population, the trait under selection must have been heritable; the degree of response to artificial selection indicates the degree of heritability. If fast growth is not heritable, it would be impossible to produce a line of fast-growing pigs by selective breeding.

The heritability of a continuously varying trait is a useful statistic in the absence of an exact knowledge of its genetics. It tells how much of the variation among individuals in a population is due to variation in their genes, which in turn tells us whether the population would respond to natural selection on the trait. (For natural selection only works on heritable traits, as we have seen.) If we are trying to understand why different animals behave differently, the value of the heritability of the behavior in question will tell us whether one possible factor—genetic variation—is at work.

Now that we have established the principles of inheritance, and the methods by which they can be studied in complex cases, let us see how well they apply to behavior. We shall consider two examples, one in which the Mendelian principle appears to apply in simple form and another in which a genetic influence on behavior has been demonstrated by artificial selection. The genetic control of behavior patterns in different species is, for all we know, so diverse that neither experiment should be thought of as particularly representative of behavior as a whole. They only show how it can be studied in some cases.

4.2   Behavioral genetics

The first example concerns the honeybee. Honeybees lives in hives in large societies of about 15,000 bees. In nature, bee hives may be built in any convenient hollow, such as the inside of an old tree. The hive consists of a series of vertical, wax honeycombs in which the bees store their food and rear their offspring. The offspring live in the combs while they are eggs, larvae, and pupae. They then emerge as adults, ready to contribute to the work of the hive. Most bee hives nowadays are made by humans, and the vertical shelves on which the bees build their combs are built so that they can be slid in and out of the hive.

Honeybee larvae are susceptible to disease. One such disease, called American foul brood, is caused by the bacterium *Bacillus larvae*. A larva that catches foul brood dies. The dead, rotting larvae may then infect other larvae in nearby combs. Some strains of bees prevent the spread of the disease by removing the rotting larvae. Such strains are called "hygienic." The hygienic bee performs two acts. It removes the cap of the cell containing a rotting larva, and it throws the larva away. Not all bee strains are hygienic; W.C. Rothenbuhler, therefore, could experiment on the inheritance of the habit and showed that the difference between hygienic and nonhygienic strains is due to their possession of different genes.

First Rothenbuhler experimentally infected combs of larvae with foul brood. He then slid these combs into different hives, and simple observation revealed which hives contained hygienic bees. He then crossed a hygienic strain with a nonhygienic strain, and tested their hybrid progeny with combs of infected larvae. All the hybrids were nonhygienic (Figure 4.3). The genes for hygienic behavior therefore must be recessive. In the next stage, Rothenbuhler did a "backcross" of the hybrids with the hygienic parental strain. He reared 29 colonies from such backcrosses in all. Six of the 29 were hygienic ($^6/_{29}$ is approximately one quarter). About another quarter ($^9/_{29}$) behaved very strangely; bees of these nine colonies took the caps off the larval cells, but then left the dead larvae inside. The remaining half ($^{14}/_{29}$) of the colonies were nonhygienic—they left cap and larva alone. Rothenbuhler thought that hygienic behavior might be controlled by two recessive genes. One gene would control uncapping, the other the throwing out of the larvae. The nine colonies that uncapped but left the dead larvae inside had one of the genes, that for uncapping, but not the other for removing. Rothenbuhler

Figure 4.3

Rothenbuhler studied the genetics of hygienic behavior in honeybees by crossing a pure nonhygienic strain with a pure hygienic strain. His results are explained by supposing that there are a pair of genes, one each of two kinds. One gene controls the uncapping of the brood chamber, the other the removal of the dead developing bee larva. If this interpretation is correct it should be possible to breed four different kinds of honeybee: fully hygienic (remove chamber caps and diseased larva); remove chamber cap but not diseased larva inside; remove larva but not cap; fully nonhygienic (remove neither). From the cross of the two pure strains, and experimental removal of caps in nonhygienic hives, Rothenbuhler found all four kinds.

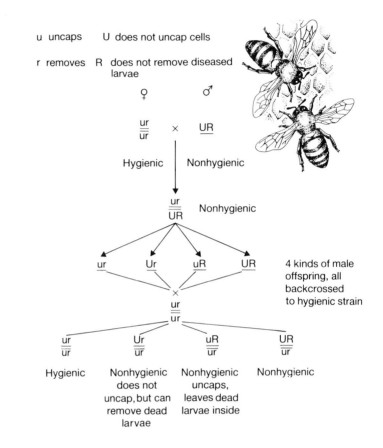

therefore reasoned that, of the 14 colonies that were apparently non-hygienic, about half (that is, about seven of them) might have the gene for removing the dead larvae. Because these seven lacked the gene for uncapping, the gene for removing would not be expressed. Accordingly, he removed the caps from the combs of infected larvae. He put these combs into his 14 hives. As he had predicted, six of the colonies promptly removed the dead larvae.

Rothenbuhler's experiments on the hygienic behavior of bees provides a very clear example of the genetic control of behavior. Each of the two behavioral stages of hygiene is controlled by one gene. It is likely that most behavior patterns are controlled by more than two—maybe by hundreds of—genes. A behavior pattern controlled by hundreds of genes does not give such clear categories as those found by Rothenbuhler. In Rothenbuhler's experiment there were two genes, with two alleles each, and four (2^2) categories of bee. If there were 100 genes, each with two alleles, there would be 2^{100} categories; the experiment would therefore be more complicated.

When animals do not fall into a few simple categories, quantitative genetic techniques become more practical. For example, the Russian geneticist Dmitry Belyaev was interested in how dogs became "domesticated"—that is, how they evolved to live with humans. During domestication, dogs became "tame": they grew less likely to attack or run away from humans. Wild dogs do not behave toward humans in the manner of domestic dogs. Now, "tameness" is a complex trait. It does not fall into simple categories, and no single Mendelian genes have been identified that influence tameness. However, Belyaev was able to study it by an artificial selection experiment. Silver foxes are fairly close relatives of the ancestors of domestic dogs, and they are also bred on special farms in Russia for their fur. Belyaev began an experiment in the 1950s in which he selected for increased tameness in silver foxes. He began with an unselected population and measured the tameness of the individual foxes in it. Tameness is not a standard property, like length that can be measured with a ruler. Belyaev had to invent a measure of tameness, and his measure was a complex score of various observable behavior patterns, including how likely the foxes were to shy away from humans, and to bite. Belyaev then bred the next generation from the tamest foxes in the population, and had continued the procedure for 18 generations by 1978 (Figure 4.4). Tameness showed an increase in the selected population. The fact that tameness responded to selection shows that there was genetic variation for tameness in the population, as each generation was bred from the foxes with genes predisposing them to behave in a tamer manner than the average for the population.

The foxes that were selected for tameness showed some interesting associated changes (that is, changes that were not specifically selected for) that are reminiscent of the behavior of domestic dogs. "Like dogs," Belyaev wrote, "these foxes seek contact with familiar persons, tend to get close to them, and lick their hands and faces. In moments of emo-

Figure 4.4

Artificial selection of tameness in silver foxes. These are frequency distributions of tameness (on the x-axis) in generations 1, 2, 10, and 18. "Tameness" was measured as a complex score of several behavior patterns, including likelihood of approach to, and of biting, nearby humans. Tameness increased as successive generations were bred from the tamer-than-average silver foxes. *(Simplified from Belyaev)*

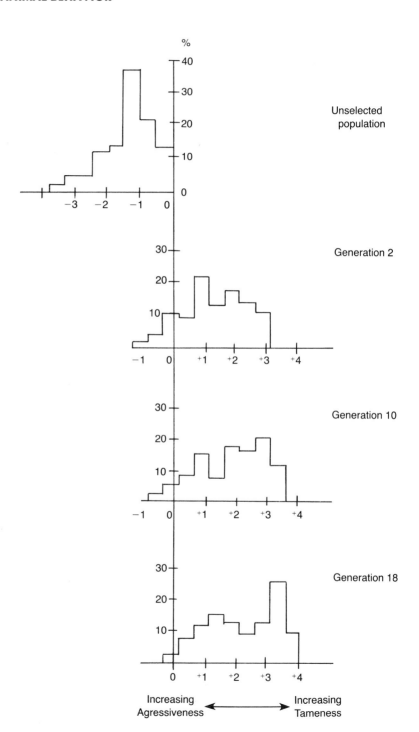

tional excitement, they even sound like dogs. There is something moving in the emotions of these foxes, that at the sight of even a strange person, they try actively to attract attention with their whining, wagging of tails, and specific movements." Barking and wagging the tail were not traits that Belyaev had selected for, but they showed up in the foxes that were selected for tameness. The set of associated changes may have been due to a common hormonal influence. Belyaev measured the levels of several hormones that operate in the hypothalamus, midbrain, and hippocampus, in the fox's brain; the neurohormone serotonin, for instance, showed higher concentrations in the selected foxes than in unselected foxes. Maybe, in selecting for tameness, he had actually bred for certain hormonal changes, and all the behavior changes characteristic of domestication arose in consequence. However that may be, the experiment does show how it is possible to study the genetics of a complex trait by artificial selection.

4.3   Development

When discussing the genetics of a trait (such as tameness in dogs or hygiene in honeybees), it is easiest to think of the trait as constant within an individual. The principles apply equally to inconstant traits, but are more clumsy to express. In fact, of course, many behavioral traits are not constant throughout the life of an individual; they are subject to developmental change. How does behavior develop? We cannot make generalizations as broad as those for inheritance, because development is less well understood. However, there have been a number of revealing studies of behavioral development that suggest some reasonably broad principles.

Many studies of development consider what factors an individual must experience in order to acquire a particular behavior pattern. Consider, for instance, the song of the male cricket: do the males learn the song by listening to other males? The crucial experiment is to rear male crickets without allowing them to hear the song of other crickets. If crickets learn their song, the experimentally isolated males should not be able to sing a cricket song. In fact they can; learning is unnecessary. Notice that the experiment only rules out one (or a few) experimental factors. It shows that the sound of other males' singing is not necessary for the development. It does not show that no experience of any sort is necessary, as indeed it could not, for it is logically impossible to substantiate universally negative statements. That, however, does not prevent us from drawing particular, limited conclusions from such "isolation" experiments.

In crickets, the ability to sing can develop without the experience of song, but in many birds the story is not the same. In most species of birds, a male reared apart from other singing males will not develop a proper song at all; this is not true of simple bird sounds, such as a cock's crow, but no songbirds are known to be able to develop their elaborate songs in isolation. If, however, the isolated bird is played a tape recorded song of its own species, it will later be able to sing it normally. (We will return to this subject.) Another case in which experience has been shown to influence development is the pecking of chicks. Gull chicks, for instance, peck at their parents' bills, which causes the parent to regurgitate food. Newborn chicks will peck at an adult gull's bill—or even a model of one—without any prior experience. Jack Hailman found that a herring gull (*Larus argentatus*) chick will initially peck equally at a model of a herring gull adult, or of a laughing gull (*L. atricilla*), which is quite different, its head being mainly black, unlike the white head of a herring gull. After a few days in the nest, however, the chicks will peck more at the herring gull model (Figure 4.5). They have experienced being fed by herring gulls, and have modified their preference accordingly, in a way

Figure 4.5
At birth, herring gull chicks will peck equally at models of adult bills of either herring or laughing gulls, but after seven days they have developed a preference for the model of the adult of their own species. Figure 7.1 (p. 166–167) contains photos of gull chicks pecking at bills of adult gulls and models. *(After Hailman)*

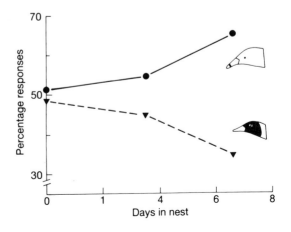

that Hailman calls "perceptual sharpening." Experience here influences development.

Actually, this experiment alone does not demonstrate that experience is necessary: the change could take place automatically with age. We should distinguish learning from "maturation." Maturation means that the change in the behavior is due to other changes in the animal, not to practice of the behavior itself. Chick pecking again provides an example, but this time we consider chicks pecking food on the ground, not at a parent's bill. As soon as a domestic hen chick hatches it starts pecking at grains that look like food, and as it grows older its aim at food grains improves. This is in part due to practice, but there is also an influence of maturation. If a chick is prevented from pecking at food during its second day, it will still be better at pecking on its third than on its first day (Figure 4.6), but it will not be as accurate as it would have been if it had been allowed to practice. Accurate pecking in chicks therefore develops, by processes both of maturation and learning, from an initially inaccurate condition to one of greater accuracy. When a behavior pat-

Figure 4.6
Chicks feed by pecking at small grains of food. Their pecking is not perfectly accurate to begin with, but improves with time. These graphs show that there are two components in the improvement. One is learning: the error rate decreases with practice—each line illustrates a decrease in the error rate since starting. But there are five lines, for different classes of chicks allowed to start to peck from one to five days after hatching. The initial error rate decreases from one to five days, which indicates that the improvement within a line is due to maturation as well as learning. Chicks started on their third day, for instance, made fewer errors to begin with than chicks started on their first day. Even without practice, the accuracy of their pecking improves. *(After Cruze)*

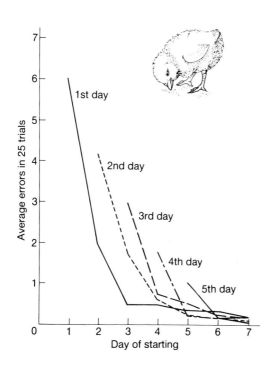

tern improves, or simply changes with age, some experiment like that in Figure 4.6 needs to be done before we conclude how much the change is due to learning from experience.

We have considered whether an experience of particular factors might be necessary for the development of given behavior patterns. Let us now make the question more precise and ask whether the experience has the same effect at any time, or whether there are particular "sensitive periods" at which it will be more influential than at others. Most research on sensitive periods has been conducted in relation to the phenomenon called imprinting; it will therefore be convenient to take the two together.

4.4   Imprinting

If you remove the eggs of a goose from their mother, you can hatch them in an incubator. When the goslings hatch, they will recognize the first moving object they see as the creature to follow around for the next few weeks. If it is a human that they see first, they will follow him or her (Figure 4.7). It helps if the human makes some distinctive noise as well: the goslings can then recognize the human by sound as well as sight. The

Figure 4.7
Goslings will "imprint" on a human foster parent if that is the individual they see after hatching. These goslings have imprinted upon Konrad Lorenz.
(Photo: Niko Tinbergen)

goslings will have become "imprinted" on the human. In nature, of course, the first moving, honking thing that a gosling sees is its parent, and it will normally become imprinted on, and follow, its mother. It is only when intrusive ethologists steal and hatch eggs that the wide tolerance of the goslings is revealed. Goslings will not only imprint themselves on human beings or geese, but also on inanimate objects such as a cardboard box, a rubber ball, or even a flashing light. Young birds, however, are not completely undiscriminating; they do have preferences. Pat Bateson has found, for instance, that inexperienced one-day-old domestic chicks prefer a red flashing light to a yellow one, if given the choice. Why they do is not clear, but the preference exists. However, if a chick is first imprinted on a yellow flashing light, it will then prefer it to a red one.

After the young animal has imprinted itself on a particular individual, its attachments are fairly irreversible. It will continue to be attached to that individual even after the period of parental care is over, not switching its attachments to some other animal. A lamb will become imprinted on whoever feeds it. If that individual is not its mother but a human foster parent feeding it from a bottle, the lamb imprints on whoever held the bottle. Even after the lamb has been weaned and has joined a flock it remembers the hand that fed it. If its human foster parent comes nearby, the lamb will leave its flock and stand near him or her.

The kind of imprinting that we have been considering so far is called "filial" imprinting, the imprinting of the response of young animals to follow their parents. Filial imprinting takes place in many species of birds and mammals. These are the kinds of animals with the most extensive parental care. Imprinting is adaptive because it enables the young to recognize and follow their parents. They will grow up in a world of many hostile enemies and one or two protective parents. If they are to survive, it is important that the young should choose the right animals to follow.

Other kinds of behavioral responses can also become imprinted. Sexual imprinting, for example, concerns the species to which the animal will direct its sexual behavior. In geese, sexual imprinting and filial imprinting are two different processes. A goose that follows a human being around as if it were its parent does not have its sexual behavior similarly disturbed; when it grows up it will court other geese. Another kind of imprinting may take place in young salmon smolt. According to one hypothesis (p. 127–129), the smell of its home stream is memorized by a young salmon, and when it grows up it will migrate back to the river that smells like its home stream. If this is so, the migratory response is imprinted in salmon.

Imprinting takes place during a defined "sensitive period." It is usually established during a specific period early in the animal's life, and the exact timing of the sensitive period differs between species. Domestic chicks, for example, only follow objects they have seen during the first three days after hatching, whereas for mallard ducklings, the phase lasts for 10–15 days after hatching. The chick will not imprint on objects seen after that time.

Sexual imprinting also occurs in early life. Most experiments on sexual imprinting have been done on birds. It has been found that birds are most easily sexually imprinted on their own species, fairly easily on closely related species, and only with difficulty on very different species. Herring gulls and lesser black-backed gulls are similar species. Black-backed gulls reared by herring gull parents (because some ethologist has moved eggs between nests) become sexually imprinted on the herring gull: the adult lesser black-backed gulls so produced will try to mate with herring gulls. They are sometimes successful. Most of the gulls that are hybrids between the two species around the British Isles are the result of these experiments. Sexual imprinting normally functions in the wild to ensure that the animal will, when it grows up, choose a mate of the correct species. In nature, looking at your parents is a good method of learning the characteristics of your own species.

4.5   Learning and memory

Learning can, in a serviceable but imperfect definition, be said to include any change in an individual's behavior that is due to its experience. A convenient distinction, which helps to organize the research on learning, is that between associative and nonassociative learning. In associative learning, the animal learns that different properties of the environment—

or different stimuli, as the properties are called—are associated, and modifies its behavioral responses to them accordingly. For example, a chimp may learn an association between poking a stick into termite mounds and extracting a stick covered with edible termites. It will then learn to poke sticks in termite mounds when it is hungry. In nonassociative learning the animal also learns to modify its behavior, but not because of any association of stimuli. Concrete discussion should make the distinction clear.

4.5.1 Nonassociative learning: habituation and sensitization

Habituation and sensitization are two of the main kinds of nonassociative learning. We can take our examples from the sea hare *Aplysia* (see p. 53). *Aplysia* breathes through its gills, which are situated in a region called the mantle cavity; the gill's enclosure opens to the outside through an opening called the siphon. If an experimenter prods the siphon, the *Aplysia* withdraws siphon and gills and folds them up within the mantle cavity. This is called the siphon (or gill) withdrawal reflex, and is simply a protective reaction. After a while, if undisturbed, the *Aplysia* puts its siphon out again, and if it is then prodded a second time, it will show the same withdrawal reflex. However, it will not do so an indefinite number of times. If it is repeatedly prodded, it comes to ignore the stimulus, and leaves its siphon and gills out. This is the kind of behavioral change called habituation: the *Aplysia* has learned not to respond to an apparently harmless stimulus. Sensitization is the opposite kind of change. Habituation means to become less sensitive to a stimulus, sensitization more so. If an *Aplysia* receives an alarming stimulus such as an electric shock on the tail, it then responds more readily to other stimuli (such as prods to the siphon) that it would otherwise have been less responsive to. It has become more sensitive. The sensitization of siphon withdrawal by a tail shock depends on a number of factors. One is whether siphon withdrawal is already habituated (i.e., the animal does not withdraw its siphon when prodded); if it is, the tail shock immediately "dishabituates" the reflex so that the siphon will now be withdrawn if prodded. The dishabituation of the siphon withdrawal reflex appears within 90 seconds of the tail shock.

The effect makes sense, because if an *Aplysia* receives a dangerous stimulus naturally, it probably means some hazardous entity is nearby, and a stimulus to the siphon is less likely to be harmless. If siphon withdrawal is not already habituated, a tail shock has little or no effect on siphon withdrawal after 90 seconds. Sensitization now appears more slowly, after a few minutes or 20–30 minutes, depending on the experiment. Unlike the immediate effect of dishabituation, the adaptive significance (if any) of the slower-acting sensitization is unclear.

The habituation and sensitization of the siphon withdrawal response in *Aplysia* are understood neurophysiologically (Figure 4.8). The nervous control of the gill withdrawal reflex is a simple unit of one sensory neuron and one motor neuron. The siphon contains the sensitive end of the sensory neuron which, at its other end, is directly connected at a synapse with a motor neuron that controls the muscles of the mantle cavity. When the sensory neuron is stimulated, it fires the motor neuron, and the siphon and gills are withdrawn. How does the system habituate? There are two possible mechanisms for regulation in so simple a system: either a change in the amount of neurotransmitter released by the sensory neuron, or a

Figure 4.8
The nervous control of siphon withdrawal in the sea hare *Aplysia*. When an *Aplysia* is tapped on the siphon, it withdraws its siphon and gills into its mantle chamber, but if it is repeatedly so tapped, it ceases to respond. This "habituation" is controlled in the synapse of the siphon's sensory neuron and motor neuron. The sensitization of siphon withdrawal by a tail shock is controlled by another neural circuit. Figure 3.6 shows a photo of live *Aplysia*.

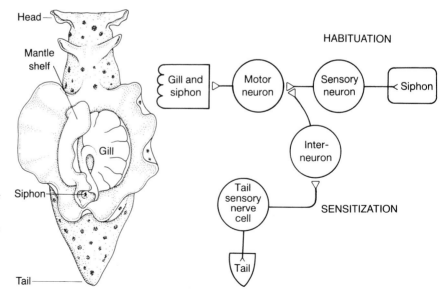

change in the sensitivity of the motor neuron to constant doses of neurotransmitter. In the case of *Aplysia* the former possibility is the real one. It is as if repeated activity in the sensory neuron exhausts its supply of neurotransmitter, making the system as a whole less responsive.

Two more kinds of neurons are needed for sensitization (Figure 4.8). The dangerous stimulus is sensed by another sensory neuron, which is connected by a synapse to one or more interneurons, which eventually connect with the synapse we met before—the one connecting with the motor neuron that controls the mantle muscles. When the sensory neuron (for instance, in the tail) becomes active, it fires the interneuron, which in turn causes a series of chemical changes in the motor neuron. The effect of those chemical changes, the details of which are known, is to make the motor neuron fire more readily. It requires less of a stimulus to become depolarized. Hence the system is sensitized.

Habituation and sensitization are not the only kinds of nonassociative learning: imprinting, which we discussed in an earlier section, is another kind, and the development of bird song, which we shall discuss later, yet another (though it may be closely related to imprinting). It has been a matter of controversy whether the nonassociative kinds of learning are really learning at all, and if so, whether they occur by the same general mechanisms as associative learning. It is likewise a matter of unsettled controversy how far the kinds of neuronal mechanisms identified in *Aplysia*'s nonassociative learning could apply to learning in general. Comparably clear facts are not available for other systems, and as we move on to associative learning we must return from the neuronal to the behavioral level. Most of the research on learning has been carried out on unnatural behavior patterns in the laboratory, particularly in two species: rats and pigeons. The kind of associative learning shown by rats and pigeons in these experiments is often called conditioning. Let us see what this means.

4.5.2 Associative learning: classical and operant conditioning

There are two main kinds of conditioning, called classical conditioning and operant conditioning. We shall take them in turn. Classical conditioning was first studied in dogs by the Russian physiologist Ivan Pavlov (1849–1936). Pavlov's interest was in digestion. In 1904 he won the Nobel Prize for his work on the physiology of digestion, after which he turned to study the conditioning of digestion. In a typical experiment Pavlov would sound a bell when bringing a dog its food. As the dog learned the association between the sound of the bell and being fed, it salivated on hearing the bell in expectation of its meal. Soon Pavlov could make his dog salivate just by sounding the bell, even without bringing its food (Figure 4.9). He claimed the salivation of the dog had been conditioned. Classical Pavlovian conditioning is advantageous to the animal, because it can respond more rapidly or appropriately to important environmental stimuli, such as those associated with food or other members of its species.

The advantage can be demonstrated by experiment. Blue gouramis (*Trichogaster trichopterus*) are fish that usually behave aggressively when encountering a conspecific in the same aquarium. In particular, males tend to attack females during the breeding season, a tendency that delays the breeding of a pair. In an experiment, Karen Hollis, Elizabeth Cadieux, and Maura Colbert trained eight males each to associate an illuminated red light with being exposed to a female in the other side of the aquarium (but behind a transparent barrier) 10 seconds later. Four other "control"

Figure 4.9

Classical conditioning of salivation in dogs. The dogs come to salivate on hearing the sound of a bell, which they associate with being fed. The trial number (on the horizontal axis) is the number of times the dog had been fed and heard an associated sound, before any measurement was made. The amount of salivation was then measured when the dog heard the bell, but without any food being brought. (*After Anrep*)

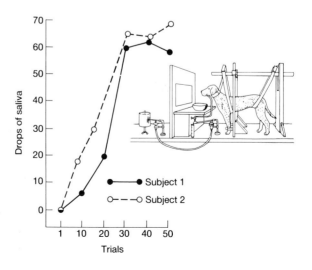

males were not so trained. The eight "trained" males soon came to perform a "frontal display" at the barrier concealing the female when the red light came on, before the female was visible. (Both sexes of blue gouramis use the frontal display when they encounter a conspecific. It consists of facing the other fish—or barrier, in the experiment—head on and holding the fins erect.) The frontal display was conditioned by the red light, just as salivation was by the bell in Pavlov's experiments. Then, after 12 days of this treatment, four of the trained males had the barrier separating them from a female removed 10 seconds after a red light was switched on; the other four trained males had their barriers removed but without a light being switched on; and the four untrained control males had their barriers removed 10 seconds after a red light was switched on. Figure 4.10 reveals the investigators' finding: the males who had learned the association of red light and seeing a female, and who were shown a red light, tended to court the females; the other males tended to attack them. Because courtship is more likely to lead to reproduction, natural selection would favor the conditioning habit.

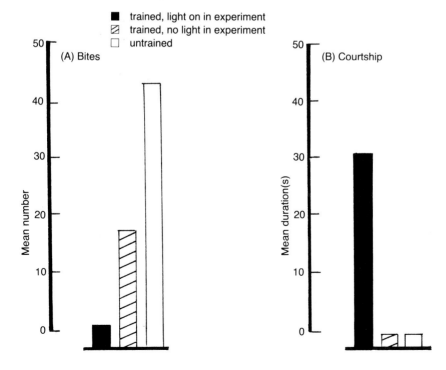

Figure 4.10

Male blue gouramis respond differently to a female according to whether they have been conditioned to associate a red light and seeing a female. In the experiment, there are three kinds of males: those who have been trained to associate a light with seeing a female and who either (a) are put with a female after seeing a light or (b) not seeing a light; and (c) control males who have not been trained to associate a light with seeing a female. The trained males who receive the light stimulus are less likely to bite the female and more likely to court her. *(After Hollis, Cadieux, and Colbert)*

The other main kind of conditioning is called "operant" or "instrumental" conditioning. It was first worked on at the turn of the century by the American psychologist, E.L. Thorndike, and later by another American psychologist, B.F. Skinner, and by many others. The experimental apparatus they used has come to be called a Skinner box. The pigeon in the box has a choice of two colored disks to peck at. If it pecks at one it receives a grain of food; if it pecks at the other it does not. Many experiments are possible. One disk may give food at one rate, and the other at a slower rate. Or, sometimes when the animal pecks the disk it may be given food, but sometimes not. The interval may be varied between the pecking and when the food is presented. In every case, the bird learns the association between doing something and being fed, and accordingly it more or less accurately pecks the correct disk when it is hungry (though pigeons do learn more rapidly in some conditions than in others). The important point is that the animal learns an association between its behavior (pecking) and a consequence of the behavior (being fed), and modifies its behavior appropriately.

Analogous experiments have demonstrated associative learning in rats. Instead of pecking at disks, rats can be taught to press levers in order to obtain food and water. They can also be taught to run through mazes. The simplest maze is a T-maze in which the rat has to make one directional choice; by rewarding the rats that turn one way rather than the other (or by giving them an electric shock if they turn in the "wrong" direction), the experimenter can teach them to make a consistent and predictable choice. The results (e.g., Figure 4.11) can be plotted on a graph to show the rate of "correct" choices in relation to the number of times the rat has run through the maze. The rate of correct choices increases with the number of trials, as the rat learns to modify its behavior (its direction of turning) because of a consequence of the behavior itself (the consequence being food or an electric shock). Again, it has learned an association.

4.5.3 How songbirds learn bird songs

Associations with bells or illuminated disks are rather artificial, though they are intended as experimental versions of things that would be im-

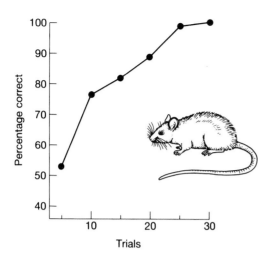

Figure 4.11
Operant conditioning in maze running in rats. The rats have to turn in a certain direction in a T-maze in order not to be electrically shocked, or in order to be fed. They choose the correct direction with increasing accuracy as the number of trials (i.e., times they have run through the maze) increases. *(After Clayton)*

portant in nature. More natural kinds of learning have also been studied directly; the development of bird song is an example.

All birds can utter a range of about a dozen different calls (such as alarm calls, p. 170), and these are all unlearned and produced by all individuals in a species. In songbirds, adult males produce another, much more complicated, kind of sound in the form of song during the breeding season. These songs have been the subject of developmental research. The pattern of learning has been studied in several species, and is probably similar in the European chaffinch (*Fringilla coelobs*) and bullfinch (*Pyrrhula pyrrhula*), the Australian zebra finch (*Poephila guttata*), and the North American white-crowned sparrow (*Zonotrichia leucophrys*). Let us consider the exact results for the white-crowned sparrow, a species that has been studied by Peter Marler and others. In the normal course of development, the chick and young bird do not sing. Then, after about two months, young male birds start to sing in a preliminary way; at that stage the sound the birds produce is called "subsong." Fernando Nottebohm described subsong as follows: "subsong is soft on volume and variable in structure and it is often done while the young bird appears to doze. Charles Darwin pointed out the similarity between subsong and the babbling of human infants; both seem to be early stages of vocal practice." At some point the variable sounds of subsong "crystallize" into a final adult song, and the form of the song changes little

(Figure 4.12a). However, if the young bird is reared in isolation for the first 50 days of its life, it never learns a recognizable white-crowned sparrow song (Figure 4.12b). This time—days 1 to 50—is the "sensitive period" for song learning, and it may be the time when the neuronal circuits that control song are developed in the bird's brain.

But it is not enough for the bird to hear its species song during its sensitive period. If it does hear song at that time, but is then deafened before it starts to sing subsong, it never learns to sing the kind of song it heard while in its nest (Figure 4.12c). It has therefore been suggested that, during their first 50 days, the white-crowned sparrows memorize the normal song of the species, and form a "template" inside their heads with which (unless deafened) they later compare their own attempts at song production. With practice, they bring their own song into line with the memorized template.

The young bird, in its sensitive period, already has a good "idea" of what song it is to remember. If it has songs of its own species and of other, related species played to it in the sensitive period, it will finally sing only the white-crowned sparrow song (Figure 4.12d). This suggests it has a preference in its memory system for its own species' song; if the young bird is isolated and has only the song of another related species played to it, it hardly does any better than when reared in silence (Figure 4.12e, compare to 4.12b). (However, it is interesting to note that if the white-crowned sparrow is reared by a song sparrow foster parent, rather than just having that species' song played at it by a tape recorder, it does learn the song sparrow's song. A live tutor is somehow a more powerful stimulus than just sound alone.)

In summary, during the sensitive period from days 1 to 50, the white-crowned sparrow memorizes the song of its own species, though it does not at that stage sing. It then starts to sing, initially in the form of crude subsong, and improves with practice until it sings a good version of its species' song. It then crystallizes its song in that form and continues to produce it through its adult life. A similar course of development has been found in the songs of other species of songbirds, though the exact timing of the various stages, and other details, differ among species.

The development of bird song is most obviously classified as nonasso-

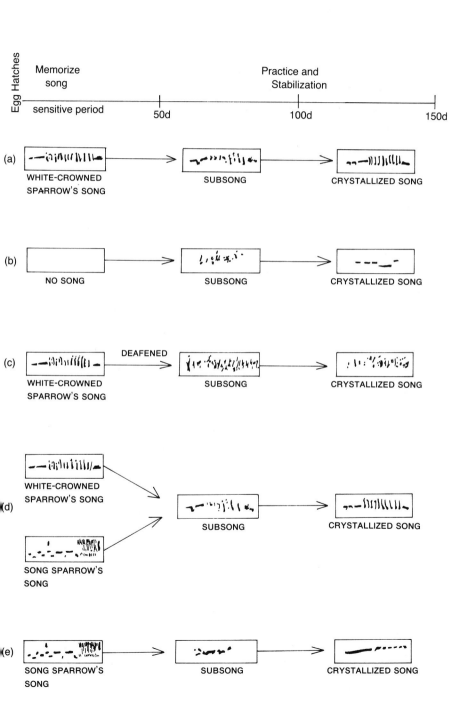

Figure 4.12

Development of song in white-crowned sparrow. The pictures are sonograms, which illustrate the frequencies of sounds being made through time. (a) Normal course of development: the bird hears its species' song while young, and initially produces a "subsong," which crystallizes into the final adult song characteristic of the species. (b) If the juvenile bird is isolated from the sound of all songs, it produces subsong but when it crystallizes it is not a proper adult song. (c) Similarly, if the young bird is allowed to hear song but is then deafened, it never improves its subsong into proper adult sing. (d) If the young bird has songs of both its own species and of a closely related species played to it, it still develops its own species' song as normal (like in (a)). (e) If the young bird is reared with the sound only of the related species, it neither learns the related species', nor its own species', song. *(Modified after Marler)* ciative learning, because the bird learns its song by comparing it with a template rather than because of any consequence of singing a better song. A better song at the learning stage does not, for instance, allow it to defend a territory or court females more effectively—although it may well have such a consequence later, after the song has been learned (p. 168).

Species vary in how flexible they are about the range of songs they will learn. The white-crowned sparrow learns only songs that are at least very like that of its own species. However, its flexibility is influenced by its rearing conditions (such as whether it grows up with an inanimate tape recording or a live foster parent) and it is difficult to be sure how much of the reported variation among species is due to the nature of the birds and how much to the varying experimental conditions. However, it is reasonably well demonstrated that birds that specialize in vocal mimicry, such as parrots and mynah birds, can learn to reproduce a much wider variety of sounds; other species, such as canaries, do not crystallize on a single song, but can learn new songs while they are adults. Learning, then, is important in the development of all songbirds, but the range of what different species learn differs considerably.

4.5.4 Memory

Memory is, by definition, essential for learning; and memory, like learning, has been studied more in the laboratory, with respect to artificial behavior patterns, than with natural behavior. It is, however, a part of many described feats of animal behavior, as we may illustrate with the Baerends's study of the digger wasp *Ammophila campestris* (Figure 4.13).

A. campestris females dig burrows in sand. After a female has dug a burrow, she closes the entrance and flies off to catch a caterpillar. She brings the dead caterpillar back to the burrow, and lays an egg on it. She then catches further caterpillars, one at a time, before finally closing the nest and leaving her offspring to develop by themselves, feeding on her provisions. Such is the cycle of any one burrow, but, the Baerends discovered, a female may work on two or three burrows at a time. Each burrow may be at a different stage. The wasp will remember exactly what to do at

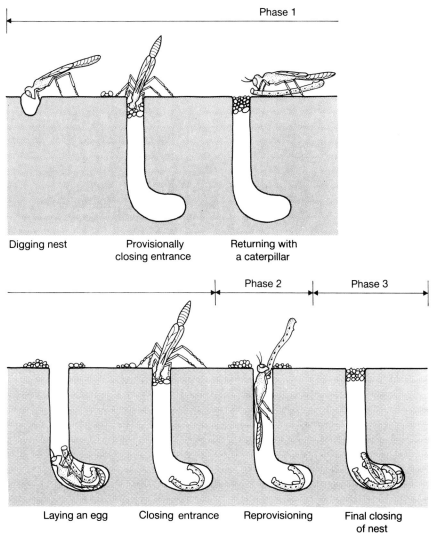

Phase 1

Digging nest   Provisionally   Returning with
                closing entrance a caterpillar

Phase 2   Phase 3

Laying an egg   Closing entrance   Reprovisioning   Final closing
                                                        of nest

Figure 4.13
The main events at the nest of a digger wasp. *(After Tinbergen)*

each burrow, according to its stage in the cycle, and the number of caterpillars it already contains, even though she may not have visited it for several days. That would not be possible without a memory faculty that can remember times, locations, numbers, and other information.

How do animals remember? What is the mechanism, or mechanisms? We can get some idea from certain experiments that have been done with rats. The experimenter first teaches his rat to perform some trick or other, such as running fast across its chamber. If the rat does not run when given some signal (such as a light coming on), the experimenter gives it an electric shock. The rat soon learns to run when signaled. The experimenter then waits for a certain amount of time and repeats the experiment to see how well the rat has remembered what it has to do. The interesting result is the relation between how well the rat remembers and how long ago it was taught its trick. There is not, as we might have expected, a simple decline in how well it remembers as time passes. If the rat is tested immediately after being taught, it does well. Rats tested up to about 12 hours later do less and less well. However, then the rat's memory seems to improve. A rat tested about 24 hours after being taught does as well as one tested just after being taught, and much better than one tested 12 hours after being taught. The experiment has been extended by trying to teach rats, at various intervals after teaching them to run to avoid a shock, to stay dead still to avoid being given a shock. One rat might be taught to run, and then taught three hours later to stand still; another rat might be taught to stay still about six hours after it was taught to run. In these experiments, the rats learn the new trick best at those times when they remember the old trick least well: it is as if a rat can learn a new trick more easily when its memory is not muddling it with the memory of the old trick.

There are two main kinds of theory of why animals forget: the decay theory and the interference theory. According to the decay theory, the memory of some event fades with time unless continually upgraded. According to the interference theory, animals forget things not because memory fades but because other memories displace them. The memory of more recent things interferes with the recall of things memorized longer ago. The rat experiments that we have been considering suggest that the interference theory is more accurate. The rat finds it easiest to memorize (learn) a new trick when the memory of the old trick is weakest.

We have now completed our main review of the positive discoveries made in the study of behavioral development. In summary, we have seen that animals modify their behavior as a consequence of their experiences; that the exact nature of the influence of experience has been studied in several systems, and the details of the mechanisms vary between systems; and that the way experience influences behavior makes sense from the point of view of fitting the animal's behavior to its environment—that is, learning is generally adaptive in nature. Now let us turn from these empirical works to a conceptual controversy that has been highly important in the history of the subject.

4.6   The instinct controversy

The word instinct is now unfashionable among the scientists who study animal behavior, but it has not always been so. Until about 1950 it was a normal part of the vocabulary. Then, in the 1950s and 1960s, it became highly controversial and it has now gone the way of all controversial terms—it is too highly charged to be useful. Although the controversy itself is now past history, it is worth knowing about, for the points of principle are highly important.

The main figure in the story is Konrad Lorenz, who began his work on animal behavior in about 1930. It was then the heyday of "stimulus-response" theories of animal behavior, according to which all behavior patterns are learned responses to associated stimuli. Lorenz's own observations led him to take a different view. He was struck by how similar the behavior of different species can be, as in (for example) the courtship of different species of ducks that grow up in very different environments. It seemed impossible to explain the similarities by similar learning experiences. He duly explained them instead by inheritance, independent of the environment. He called behavior that (he believed) develops independently of experience instinctive. Of course he did not deny that experience does influence the development of some behavior patterns, and he accordingly divided behavior patterns into instinctive (inherited patterns that develop independently of experience) and learned (the opposite). He also suggested a method to distinguish to which category any given behavior pattern belonged; this was the isolation experiment. We have met isolation experiments before (p. 89–90), but allowed them then only a narrower purpose than did Lorenz. In an isolation experiment, animals are isolated from their normal environment at birth; they have no opportunity of normal learning. According to Lorenz, if a behavior pattern developed normally under such circumstances, it belonged in the instinctive, rather than the learned, category of behavior.

Lorenz's argument might seem to move from a justified limited conclusion to an unjustified general one. An isolation experiment cannot eliminate all possible environmental influences, all sources of experience; it cannot prove a universal negative, if only because universal negatives cannot be proved. Strictly speaking, it cannot be shown that a behavior pattern develops independently of experience; only specific, identified factors may be ruled out. Indeed, in a sense, behavior cannot develop independently of the environment, and cannot, in a developmental sense, be called "inherited" at all. Behavior is part of the phenotype of an organism, not its genotype; behavior develops by an interaction of genes and environment. An account of the development of behavior within an individual would have to mention a series of environmental influences on gene expression. One cannot refer to a behavior pattern as inherited (or instinctive) or learned; the terms can only properly be used to refer to the causes of *differences* between individuals. If one hive of bees catches foul brood, but another does not, we can legitimately ask whether the difference is genetic or environmental in origin, but that is quite different from asking whether a behavior itself, in its development, is genetic or environmental. It refers only to the cause of a difference between individuals at a certain life stage, not to a whole course of development within an individual.

To identify the cause of a difference as genetic or environmental in origin, although it is a clear dichotomy, does not save the original Lorenzian dichotomy. Whether a difference between two animals is genetic or environmental is a separate question from whether experience influences the development of behavior within the two animals. It is logically possible that the difference between the behavior of two types of animal could be genetically caused, but that both types could learn the behavior in question. This would be a case of a genetic influence on learning, of which several real examples are known. Some genetically different strains of rats, for example, differ in their ability to learn to run through mazes. Likewise, a difference may be environmentally caused, but the two types of behavior not be learned. There are two types of

locust, for example, the solitary and migratory types, that are so different in form and behavior that they were once mistakenly classified as
different species, but the difference is environmentally triggered (by the
degree of crowding) and the two types are not genetically different.

All these points have been repeatedly raised against the original
Lorenzian dichotomy. Daniel Lehrman is perhaps the best known critic,
but there were (and are) many more. Lorenz responded by saying that he
did not hold the position they were attacking. On the first page of his book
Evolution and Modification of Behavior (1965) he wrote "no biologist in
his right senses will forget that the blueprint contained in the genome
requires innumerable environmental factors in order to be realized in the
phenogeny (development) of structures and functions." He went on to
clarify his distinction. In dividing behavior into inherited and learned, he
was not so much trying to explain how behavior develops as how it comes
to be adapted. He agreed that behavior is not inherited, but what he called
the "adaptive information" of behavior is inherited. By "information"
Lorenz means that, in order for an animal's behavior to be adapted to the
environment, the animal must have some knowledge of (information
about) the environment: if an animal is to perform some adaptive behavior, such as finding food, it must know what food looks like.

Now, according to Lorenz, the dichotomous distinction between instinctive and learned behavior cannot be made for development, but it
can for adaptation. The adaptation of any piece of behavior to the
environment could have either of two sources. It could be due to natural
selection over the generations, or it could be learned. A bird pecking at
food grains could have known without learning what food looks like, or
it could have learned it. In this case, as we have seen (p. 91), there is a bit
of both, but the dichotomy is not logically false as it was when applied to
development. Adaptive information can be said to be inherited, if natural selection has built that information into the animal's genes over the
generations.

Whether Lorenz's critics were burning a straw man is unimportant.
Only the conceptual conclusions really matter. Instinct has become unfashionable because of its association with an erroneous theory of development. Behavior cannot be developmentally divided into the inherited

and the learned (or environmental); all behavior is influenced by both factors. Only differences between classes of individuals can in principle be attributed to a single cause. Lorenz's point, however, that the adaptiveness of behavior is due to inherited information, remains both valid and important.

4.7   Culture in animals

The distinctive property of cultural behavior, as ethologists use the term, is the way it is passed on from one generation to the next. Instead of being inherited by the process of Mendelian genetics, it is "inherited" by imitation. An animal acquires the behavior pattern by imitating it from another. Therefore, all that is necessary for a species to acquire a culture is that its members should be capable of learning and memory, and meet other members of their own species sufficiently often to be able to learn things from them. Cultural behavior is therefore most likely to be found in species that form social groups.

The clearest examples of cultural behavior do indeed come from a social animal, the Japanese macaque (*Macaca fuscata*), a monkey that inhabits the forests of various islands around Japan. Since 1950 Japanese ethologists have been watching troops of the Japanese macaque and recording the lives of individuals within the troops. In 1952, they started leaving sweet potatoes on the beach at the forest edge, as food for the macaques, which duly came out of the forest and ate them. Next year something new was seen. Macaques were picking up potatoes, taking them to the sea, and washing the sand off the potatoes; they used one hand to dip the potato into the sea and the other to brush the sand off. Potato washing was an entirely new pattern of behavior; no macaque had ever seen a potato washed before. The habit was invented by a single two-year-old female called "Imo," who, in a moment of inspiration, had first tried washing the potatoes. Soon other macaques in her troop imitated her and the habit became more widespread. The first macaques to copy Imo were those of her own age; older macaques learned the trick later. Thus, five years after Imo's invention, 80% of the macaques in the troop between the ages of two and seven washed potatoes, but only 18%

of those eight years or more washed potatoes. There are two reasons why the habit spread faster among the younger macaques. One is that they are more willing to explore new skills; the other is that Imo herself was young, and macaques interact most in their troop with other individuals of a similar age to themselves.

Imo's career as an inventor did not end with potato washing. Two years later she surpassed this. As well as leaving potatoes, the ethologists had also scattered grains of wheat on the beach. The macaques initially picked out single grains, one at a time, from the sand; but Imo learned to pick up whole handfuls of sand and grain mixed up, and throw the whole lot into the sea. The sand sank and the wheat floated, and Imo could then skim a clean meal of wheat off the surface. Wheat skimming is more clever than potato washing. Washing a potato is only a small development from brushing earth off it, which is something all macaques do anyway; but the separation of sand from wheat required Imo to throw away the food after she had picked it up, and then wait for the sand to sink before she collected her food up again. Once learned, it is a much better way of collecting wheat. Like potato washing, it was soon imitated by other macaques in the troop. The first to learn were again those of a similar age to Imo. From them, it passed into the culture of the troop.

The habit of opening milk bottle tops probably spread through several species of birds by an analogous cultural process. The habit is best known in British tits, particularly great tits (*Parus major*) and blue tits (*P. caeruleus*), but it is also found in other species. It was first described near Southampton, England in 1921, where birds were seen removing the tops of milk bottles and drinking the milk beneath. Through the 1930s and 1940s the habit quickly spread through Britain, at far too fast a rate for it to have taken place by the natural selection of Mendelian genetics (Figure 4.14). J. Fisher and R.A. Hinde, who first collected the reports of milk bottle opening by birds, therefore suggested that the habit might have spread by imitation. After a tit had seen another tit open a milk bottle top it might then try out the behavior pattern for itself; it would discover the reward, and go on to open other bottles. Thus the habit would spread.

Figure 4.14
The spread of the habit, in tits, of opening milk bottle tops in Britain from the first record in 1921 near Southampton, until 1947. *(After Fisher J, Hinde RA. The opening of milk bottles by birds. Brit Birds 1949;10:337–403. Photos by V. L. Breeze)*

Imitation need not have been the only process at work. From the maps of the spread of the behavior (Figure 4.14) it looks as if the habit were invented independently more than once. A recent experiment, however, confirms that birds can learn the habit by imitation, and also adds another side to the story. D.F. Sherry and B.G. Galef experimented on the black-backed chickadees (*Parus atricapillus*) of Canada (these are close relatives of the British tits in which the spread of the habit was recorded). Their experiment was as follows. They first presented small milk containers "of the type often provided with coffee in restaurants," with intact tops, to 16 birds. Of the 16, 4 opened the containers spontaneously, the other 12 did not. These 12 made the main experimental subjects.

They divided the 12 into three groups of 4 birds each. The birds of one group were put, each with an unopened milk container, in one compartment of a cage that was divided into two compartments by a wire mesh through which the birds could see. In the other compartment Sherry and Galef put a "tutor" bird—one of the four that already knew how to open the milk containers—and a milk container. The bird in this treatment would have the opportunity to learn the skill by imitation. The birds of a second group were put alone in a cage with a full milk container that had already been opened by the experimenters. They were treated in this way because birds might be able to acquire the skill simply by coming across an already opened milk bottle. On seeing it, the bird might drink out of it. The bird might then learn that a certain movement of its bill in relation to a milk bottle results in a meal, and if the bird were then to perform the same activity on an unopened bottle it might break through the top for itself. In this case, the spread of the habit would still be by a nongenetic learning process, but not by imitation. Sherry and Galef's third treatment was a control: the birds were put alone, each with an unopened milk container, as in the first part of the experiment. In the final stage, the 12 birds from all three treatments were retested for their ability (when alone) to open one of the milk containers. The result (Table 4.1) was that three of the four tutored birds had learned to open the milk containers, as had three of the four that were given an already opened one, but none of the controls had learned the skill.

Sherry and Galef's experiment supports the original hypothesis, that

TABLE 4.1 Sherry and Galef's experiment on the learning of black-backed chickadees to open milk containers. The text explains the three stages of the experiment. The three birds in the "opened" treatment that drank the milk in the training session did not have to open the container. The one bird in the control treatment that opened its container in the training session did not drink the milk. The difference between experimental and control treatments is statistically significant.

	N	*Number of birds opening milk containers in:*		
		Pre-training	*Training*	*Testing*
Spontaneous opening	16	4	—	—
Treatment				
Tutored	4	0	3	3
Opened	4	0	3	3
Control	4	0	1	0

(Slightly simplified from Sherry DF, Galef BG. Cultural transmission without imitation milk bottle opening by birds. Anim Behav 1984;32:937–938.)

the spread of milk bottle opening was partially driven culturally, by imitation. But it suggests that this was not the only process. Some birds probably learned the skill after encountering already opened bottles, from which not all the accessible milk had been drunk by the original opener. The relative importance of the three processes by which the habit could have spread (spontaneous discovery, imitation, and learning from already opened bottles) is more difficult to assess. However, the purely cultural process did operate in the experiment, and therefore probably occurred in nature as well.

Cultural inheritance is analogous to genetic inheritance and leads to a process of evolutionary change analogous to genetic evolution. Darwinian genetic evolution takes place because genes are passed on from one generation to the next, and if some genes build better bodies than others, they are favored by natural selection and become more common, and evolutionary change will take place. With cultural behavior, an analogous process will operate. If one behavior is more readily imitated than another, it will become more common in the population, and evolutionary change will take place. The two processes are only analogous, not identical, and there are important differences between them. One is their relative rates. Genetic evolution is necessarily slow, because genes are only passed on once per generation; gene frequencies can therefore only change once per generation. Cultural change can be much faster. Learning by imitation can take place in a few minutes, whereas the reproduction of a whole new adult animal can take many years. There is time for many cultural evolutionary events in each generation. Genetic evolution is therefore typically a slower process than cultural change. The relative rates of the two processes have an important consequence for our understanding of humans. In that most cultivated of species, cultural changes proceed at an exceptionally high rate, and affect nearly all components of behavior. We therefore might expect cultural changes to have greatly outstripped our slowly changing evolutionary heritage, in which case our present behavior should be attributed more to the process of cultural, than of genetic, evolution. This is the reason why many ethologists are hesitant to apply the insights of the theory of natural selection, which have been gained for the social behavior of animals with only rudimentary cultures, to the behavior of our own species. Our culture may have rendered natural selection relatively irrelevant.

4.8   Summary

1. The inheritance of behavior is effected by means of molecules, called genes. The simplest scheme of inheritance has one pair of genes controlling one pair of traits (such as "tall" and "short"); a cross between a tall and a short individual will produce tall and short offspring in a 3:1 ratio.

2. One behavioral example with simple genetic inheritance concerns hygienic behavior in honeybees.

3. When inheritance is controlled by many genes, the techniques of quantitative genetics are used. Artificial selection has demonstrated a genetic influence on tameness in dogs.

4. The factors needed for the normal development of a behavior pattern can be studied by isolating young animals from hypothetically necessary factors. It has thus been shown that both practice and "maturation" influence the accuracy of pecking in chicks.

5. Some factors influence development only during a confined "sensitive period." Imprinting usually takes place in the first few days after birth.

6. Learning may be nonassociative or associative. Habituation and sensitization, two forms of nonassociative learning, are shown by *Aplysia,* and the neurophysiological control is understood. Habituation is caused by decreased release of neurotransmitter from the sensory neuron, and sensitization by release of a chemical from an interneuron that renders the motor neuron more easily depolarized.

7. Associative learning, often called conditioning, has been studied experimentally in two forms. In classical conditioning, illustrated by Pavlov's dogs and by courting gouramis, an animal learns to anticipate a future stimulus by an associated prior stimulus. In operant conditioning, illustrated by pigeons learning to peck at disks in order to be fed, an animal learns to perform one behavior pattern in order to achieve some goal such as obtaining food.

8. The learning of song in songbirds such as the white-crowned sparrow proceeds in two phases, a memory phase and a practice one. The bird first sings a crude song, and gradually brings it into line with its memorized version of its species' characteristic song.

9. Behavior patterns cannot be developmentally divided into inherited ("instinctive") and learned behavior, because the development of every behavior pattern is influenced by both factors. Behavior itself is not inherited. The "adaptive information" manifested in behavior can, however, be described as inherited.

10. A species capable of learning can develop a culture consisting of behavior patterns passed among individuals by imitation. Japanese macaques have in this way acquired a cultural repertoire of feeding techniques, potato washing, and wheat skimming. Tits have likewise learned to open milk bottle tops, in part at least, by imitating other birds. Cultures can evolve in a manner analogous to genetic

evolution. Cultural evolution will probably be much the faster process of the two.

4.9   Further reading

Konner (1982) describes various examples of genetic influences on behavior. Rusak et al. (1990) describe a short-term environmental influence on gene expression. Hailman (1969) describes his work on gulls. Gould and Marler (1987) discuss the development of bird song and Nottebohm (1989) describes the neurophysiology, including the remarkable discovery that new nerves develop in adult canaries in the brain region controlling song. For another example of the influence of experience on brain structure, in honeybees in this case, see Withers, Fahrbach, and Robinson (1993). Turkkan (1989) discusses modern work on conditioning, and Roper (1983) and Staddon (1983) introduce the biological perspective on learning; see Hollis, Cadieux, and Colbert (1989) and Hollis (1990) for experiments on the advantage of Pavlovian conditioning. On culture, see King (1991), Galef (1992), and Tomasello, Kruger, and Ratner (1993). A special issue of the *Philosophical Transactions of the Royal Society of London* (series B, vol. 329, pp. 97–227, 1990) includes articles on a number of themes in this chapter, such as one by Fitzgerald et al. on *Aplysia,* and two on memory in food-hoarding birds. The issue has been published separately as a book (Krebs and Horn, 1991). See also Shettleworth (1983) on avian food hoarding. Several chapters in Section II of Bateson (ed., 1991) are also relevant.

CHAPTER 5

Movements and

Migration

This chapter considers the mechanisms—particularly the sensory cues—by which animals find their way when migrating. We look at the problem in three stages: movement in a familiar, local environment; orientation by a compass; and "true navigation," in which an animal can navigate from arbitrary and unfamiliar starting points. The main example of true navigation—pigeon homing—is one of the great unanswered questions of animal behavior.

5.1 The principles of migration

We shall use the term migration here to refer to almost any pattern of movements by living things: the mass movement of herds of wildebeest across the plains of Africa, the seasonal migrations of birds such as swallows and of butterflies such as the monarch, the return of salmon to their natal stream. All invite two questions: why do the animals migrate and how do they know the way? Taking the first question first, an answer can at least be given in abstract terms. Each kind of animal lives best in a particular environment. The right kind of food should be available, the temperature and habitat must be right, and there must not be too many other animals that would parasitize it, or kill it for food. In some cases, the environment is so constant, or the animal can live in such a wide range of environments, that once an animal has found a suitable place to live in, it will not need to move far to satisfy all its bodily wants. Snails of the species *Cepaea nemoralis*, for example, usually die less than a hundred yards away from where they hatched. However, environments typically vary in time and space, and even if the conditions are good where the animal is now, they may not be in a month's time; it may then be better to be a hundred miles south. The abstract reason why species migrate is that environments change and the best place to be varies with time. The pattern of animal movements should follow the pattern of environmental change.

The answers to the second question, of how animals find their way when migrating, can be more varied. If the animal has a sufficient sensory range, it need only move toward areas that its sense organs reveal to offer better conditions. This kind of "planned migration" is performed by wildebeest (Figure 5.1). Wildebeest (discussed on p. 141) inhabit the plains of Africa and are frequently on the move. But they do not move blindly in the hope of coming to better pastures. Wildebeest eat grass, and grass grows after rain; they can sense where rain is falling by using their eyes and ears (but not their noses): they then migrate in that direction. Wildebeest movements follow, at a distance of a few days, the pattern of rainfall, but they only keep moving so long as rain is falling or has recently fallen within the area scanned by their senses.

Planned migration is well suited to the capricious pattern of local rainfall in East Africa. But other environmental changes are more predictable. The animal can anticipate them, rather than waiting for direct evidence. In temperate regions there are regular seasonal cycles, ultimately driven by the regular cycling of the earth around the sun. Associated with the cycle of day length and temperature are many other cyclic changes that matter to many animals: cycles of plant abundance, of leaves on trees, of the insects that live on plants. It is not surprising, therefore, that many animals perform regular seasonal migrations, northward in the spring and southward in the fall (in the northern hemisphere). The insects (mainly Lepidoptera) and birds that do so cannot directly sense the superior environment of the north (in the spring), but

Figure 5.1
Great herds of wilde-
beest migrate around
the Serengeti Plains
of Kenya, in East
Africa. Their
migrations are
directed toward areas
of recent rainfall.
(Photo: Heather Angel)

its superiority, on average, is guaranteed by the predictable changes of
the seasons.

Seasonal migrators need an accurate cue to anticipate the seasons, and
at least a compass sense to guide their movements. They measure the
seasons by the changing day length, as may be demonstrated by keeping
migratory birds under artificial day length conditions. They then migrate
at a time dictated by their experienced day length, rather than the time of
year. The effect of day length on behavior is mediated hormonally; declin-

ing production of sex hormones at the end of the breeding season prepares the birds to migrate in the fall; they do not migrate if injected with sex hormones. Similarly, the release of sex hormones in the spring, which stimulates reproduction (p. 72), also stimulates northward migration. This association led the biologist J.B.S. Haldane to remark semi-humorously that although "we must be very careful in attributing human motives to animals, the emotion behind migration to breeding places is almost certainly more like human love than hunger or curiosity." Seasonal migrators could in principle orient themselves by a simple sense of direction; all they need to do is obey some such rule as "fly two hundred miles south" or a more complicated series of directions. They could obtain compass information from the sun and stars, and it is indeed known that some seasonally migrating birds follow stellar patterns; stellar orientation has been particularly well studied in the North American indigo bunting (*Passerina cyanea*). Another important factor is that inexperienced birds can navigate simply by following adults who have learned the route. Most wild fowl navigate in this manner, though it is not a universal mechanism. Cuckoos (*Cuculus canorus*), for instance, migrate as individuals; inexperienced cuckoos have to perform their seasonal return migration without learning or imitation.

Most insects are not powerful enough fliers to be able to carry out seasonal return migrations, but a few can. The monarch butterfly (*Danaus plexippus*) is one. It is a large insect, strikingly colored in black and gold, that mainly lives in North America. In the summer the monarch is distributed from Mexico to Canada, but with falling temperatures it moves south (Figure 5.2). It has been known for a long time that monarchs can be found in winter in the southern United States, but large communes of overwintering monarchs have more recently also been found in Mexico. In the spring they migrate north again. Not all monarchs migrate south in the fall; some hibernate in the north. Those that do migrate move at astonishingly high speeds: tagged individuals move at an average of 80 miles a day, and some move much faster. If the winter is warm in Mexico they live as free individuals as in the summer, but when it turns cold they aggregate in dense groups, and remain still, in order to conserve energy. For a cold-blooded insect, the main reason for

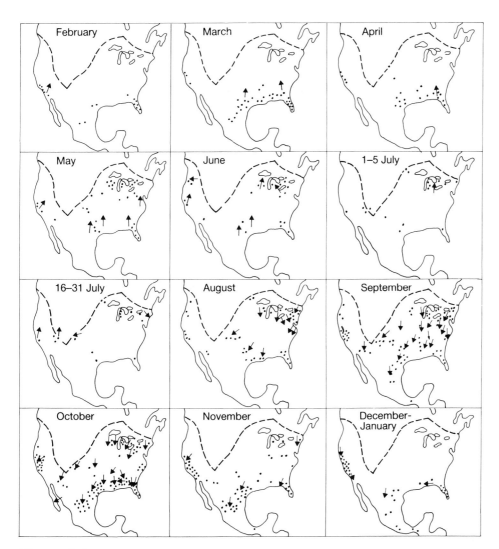

Figure 5.2   Range maps of monarch butterflies through the year. In the fall they migrate from the United States and Canada southward toward Mexico. In the spring they fly north again. It is an example of a "seasonal return migration." *(After Baker)*

seasonal migration is probably the need to be in a warm enough place to allow an active life; a warm-blooded bird does not have that problem.

Migration can have important ecological consequences, by regulating the local density of animals. If the density of animals in an area increases to the point of overcrowding, there will be insufficient food to go around, and it might pay an animal to move away. If competition is too great in one place, it may be better to try another. For this reason, an increase in population density often precipitates a round of emigration. The migrations of lemmings are stimulated by overcrowding; so too are the movements of aphids (aphids are insects that live on the sap of plants—greenfly are common examples). In most species of aphids, an individual may grow up either as a wingless stationary form or a winged migratory form; they are more likely to grow up with wings if the local population density is high.

The regulation of population density can only be a consequence of migration, not the reason why natural selection causes the habit to evolve. Natural selection only favors habits that make organisms leave more offspring; the advantage of a habit must therefore be in the short term. Population regulation, however, if it has any advantage at all, can only have a long-term one; it must therefore be a consequence of individual decisions to emigrate, taken on the grounds that conditions will be better elsewhere, not on the grounds that the population level must be kept down in order for the local resources to be conserved. Natural selection takes no account of long-term considerations.

5.2   Homing: local landmarks and home cues

The life cycles of many animals require them to find their way back to a particular place. The task for a green turtle (Figure 5.3) to find a tiny island in the expanse of the Atlantic Ocean, or a salmon to find the exact river tributary in which it was born, appears to us exceedingly difficult. Because the purpose of homing is often self-evident (it might, for instance, be a matter of finding the right place to lay eggs), the question of how they find their way is usually more interesting. We might distinguish three possible answers. One is that animals memorize local landmarks and directions on their way out, and simply reverse the directions to find their way home; a second is that the home site itself has some property that can be recognized at a distance. The third explanation, which is most likely to be important in long-distance homing, is that the animal has an internal "map" sense, and can both estimate its own map refer-

Figure 5.3
Green turtles
(*Chelonia*) live most
of their lives in the
sea, but return to
their natal island to
breed. They deposit
their eggs in the sand
on the beach.
Different populations
probably return to
different islands; the
most famous such
island is Ascension
Island, in the Atlantic
Ocean. This green
turtle female has
homed to Aldabra
Island in the Indian
Ocean. *(Photo: Tim
Guilford)*

ence and that of the home site. We can consider reasonably clear-cut examples of the use of local landmarks and of home stimuli, but when we come on to a possible map sense we shall move into one of the more unsettled areas of behavioral science.

Let us consider first the homing problem of the bee-killing digger wasp *Philanthus triangulum*. Female wasps of this species dig burrows in the sand to provide a nursery for their offspring. Each female digs several (about six or seven) cells off the side of her burrow, and lays an egg in each cell. When the egg hatches into a larva it will need food. That is where the bee-killing part of the wasp's name comes in. The mother wasp goes out of her burrow, catches a bee, stings it to death, and brings it back to the burrow. She then opens the entrance to the burrow and takes the bee down to one of the cells. She continues to catch and bring back bees, one at a time, until each larva has about two bees to eat. The mother wasp, therefore, does not merely dig a burrow and later leave it never to return: she departs from and comes back to it many times. On every return she has to find her burrow, distinguishing it from its surroundings. It is not an easy task to find a particular burrow, because these wasps can nest in quite dense groups; there might be more than twenty burrows within a circle of a five-yard radius. The problem of how the digger wasp locates her home has a special place in the history of ethology: it was one of the first questions about a behavioral mechanism ever to be asked, and experimentally answered. It was studied by Niko Tinbergen in 1929, on the heaths and sand dunes of Hulshort in Holland.

Tinbergen first confirmed, by individually marking all the nests and wasps in a particular area, that each wasp does indeed always return to her own burrow. He then performed some simple experiments to test whether the digger wasp recognized her own entrance by the distinctive array of odd objects (haphazardly fallen sticks, pine cones, stones, etc.) around it, or by some stimulus emanating from the entrance itself. He placed around the entrance of each of several chosen burrows a neat circle of pine cones. He left them there for a few days, checking that the wasps still kept returning to their burrows. He then waited for the wasps to leave on bee-hunting expeditions, and while they were away he moved the cones a yard or so away from the entrances (Figure 5.4). When the wasps returned they landed where the entrances "should" have been, in the center of the circle of pine cones. Evidently they recognized the entrance by its surrounding landmarks, not by any stimulus from the entrance itself.

Learned local landmarks are a feasible means to navigation in a local, familiar area. But longer distance homing must require other techniques. Let us now consider the techniques of the salmon. All species of salmon have similar life cycles. They are laid as eggs in river tributaries all over the northern hemisphere and they live their first few months in the river, then migrate downstream to the sea. They spend two or three years in feeding and growing out at sea, during which time they cover thousands

Figure 5.4
The bee-killing digger wasp *Philanthus* uses physical landmarks to recognize its home nest. In an experiment, Tinbergen allowed wasps to become accustomed to a circle of pine cones around its nest entrance; he then moved the circle while the wasp was away, and on its return the wasp sought the nest entrance in the circle of pine cones as before. *(After Tinbergen)*

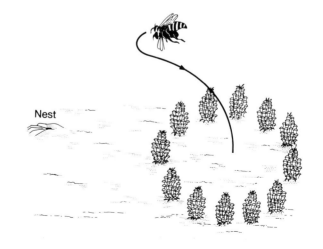

of miles; and when mature, they migrate back to exactly the same river, and same tributary, as they were born in. How do they find their way home? The answer is thought to be mainly by smell. The attractive odor may come from either (or both) of two sources: the young salmon that are still in the stream, not yet having migrated to the sea, and any other characteristic odors in the stream. Salmon have been shown to be capable of the necessary olfactory discrimination, but the most direct evidence that they use their sense of smell comes from experiments, of the kind first performed by W.J. Wisby and A.D. Hasler, in which the salmon's olfactory sense was impaired. Wisby and Hasler actually plugged the noses of their salmon, which then homed less accurately than untreated controls (Table 5.1); in more recent experiments, the same result has been obtained by cutting the salmon's olfactory nerves.

There is evidence that salmon imprint on the smell of their river during a sensitive period just before they migrate downstream. Such evidence comes from experiments in which young salmon have been transferred from their stream of birth and released in another stream. In one such experiment, L.R. Donaldson and G.E. Allen took 72,000 young salmon at the "fingerling" stage (when they are about one year old) from the Soos Creek Hatchery in Washington State (for locations see Figure 5.5) and divided them into two groups. In 1952 they released

TABLE 5.1 Numbers of coho salmon released and later recaptured in two rivers in Washington State, with and without their olfactory sense impaired. Salmon with their noses plugged homed less accurately. Figure 5.5 shows the locations. *(From Harden-Jones FR. Fish Migration. © 1968, London: Edward Arnold, using data of Wisby and Hasler)*

Stream of origin		Number released	Number recaptured Issaquah	East Fork
Issaquah	Controls	121	46	0
	Nose plugged	145	39	12
East Fork	Controls	38	8	19
	Nose plugged	38	16	3

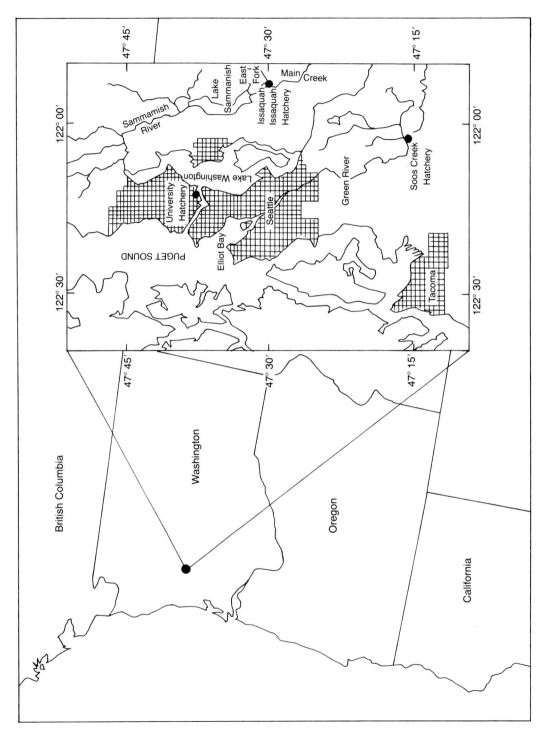

Figure 5.5   The locations of Issaquah Creek and the University Hatchery are shown on this map of part of Washington State. Experiments on the use of olfaction in homing by coho salmon were done here. *(After Harden-Jones)*

one group (identified by the removal of the right pelvic fin) at Issaquah Hatchery, and the other (which had the left pelvic fin removed) at the University of Washington Hatchery, at Seattle. The salmon returned in the winter of 1953–1954; they homed on their release sites. None of them returned to Soos Creek, the site of their birth and first year of life. Of 71 marked salmon caught at Issaquah Creek, 70 lacked their right pelvic fins; all 124 marked salmon caught at the University Hatchery lacked their left pelvic fins. Results like that suggest that the young salmon learn the odor of their native stream (or of the young salmon in it), and later find their way home by seeking that scent. But there is another explanation. Salmon migrate in schools. The results of the transplantation experiment may only be due to the transplanted salmon following the lead of the native salmon. Recent commentaries, therefore, such as the review by O.B. Stabell, maintain that the case for imprinting is at best not proved. But it is not in doubt that salmon rely critically on their sense of smell to guide them home. Whether they learn their home stream's distinctive smell during a discrete sensitive period is undecided.

5.3 Pigeon homing

It is less easy to believe that the large-scale migrations of birds are accomplished by memorized landmarks and home cues than are those of the digger wasp and salmon. Moreover, experiments on pigeons have been thought to rule out that possibility. Pigeons taken from their home loft to some unfamiliar place, say one hundred miles to the east, and then released, are capable of finding their way home. After its release a pigeon circles a few times around the release site for a few minutes and then flies off, usually in the approximate direction of its home loft. A few hours later it will arrive home. Then, a few days later, after the pigeons have had time to rest, they may be taken to some other point, perhaps one hundred miles south this time, and they will repeat the trick. Homing pigeons therefore appear to be able to find their way home from any starting point (provided it does not exceed their maximum flight distance), whether or not they have seen their starting point before. We shall call this ability "true navigation": the ability to find the way through unfamiliar areas.

True navigation is not just a matter of finding the way home from a familiar starting point, which can be achieved (as by the digger wasp) with the use of memorized landmarks and learned orientations to them. Nor can true navigation be achieved only by "compass orientation." A compass indicates a direction and compass orientation means moving in a set direction to a compass. For example, the animal might always fly southward; regardless of starting point it would need some internal "compass" sense but nothing more. Compass orientation can explain the seasonal movements of birds and butterflies: they could find their way simply by flying in a particular direction. They probably possess more sophisticated navigational equipment, but seasonal movements alone do not demand such skills. The test between compass orientation and true navigation is to move an animal experimentally away from its normal tracks. If it just flies by the compass, it will show an equivalent displacement from its goal; if it can navigate it will reach its goal despite the displacement.

This type of experiment has been done with starlings in Europe. Starlings migrate during the fall from the area around the Baltic Sea to their wintering places in southern England, northern France, and Holland. One year, some ethologists caught and marked (with rings around the

Figure 5.6

In northern Europe, starlings normally migrate southwest during the fall; the points on the left-hand map show where starlings that were banded in The Hague, Holland, were later recovered in the winter and in the breeding season. On the right are the results of an experiment in which starlings caught over Holland in the fall were transported to and released in Switzerland. Juveniles continued to fly southwest, but adults adjusted their direction to take them to their normal winter sites.

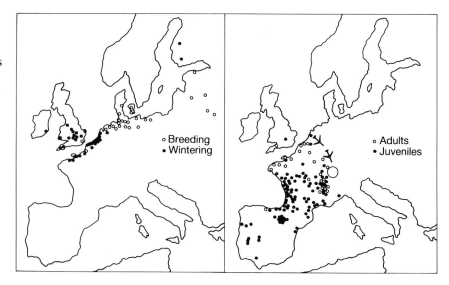

birds' legs) a number of starlings as they were flying over Holland, put them in some airplanes, flew them to Switzerland, and released them. The juvenile starlings now behaved differently from the adults. The juveniles continued to fly southwest, and wintered in southern France and Spain. The adult starlings, however, flew northwest to the usual overwintering grounds (Figure 5.6). Evidently the young starlings used only compass orientation whereas the adults used true navigation. The same kind of compass orientation is shown by adult mallard ducks in the phenomenon called "nonsense" orientation. When mallards are taken away from Slimbridge, in southern England, whatever the direction they are taken away they initially fly northwest, even if it is the opposite direction from home (Figure 5.7).

The bird for which there is the most evidence of a full navigatory sense is the pigeon. Humans have artificially selected (p. 83) pigeons for their homing ability for thousands of years; pigeons were used to carry

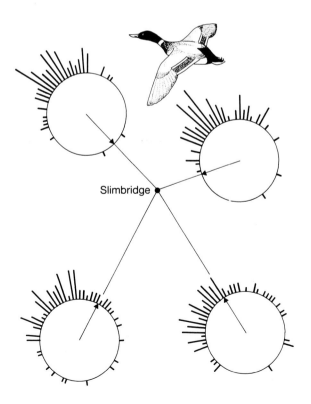

Figure 5.7
"Nonsense" orientation of a mallard duck. Whatever direction mallards are taken from Slimbridge, in England, they fly towards the northwest after release. The arrow inside each circle indicates the correct return polarity. The lines indicate the numbers of ducks flying off in that direction from the release site. *(After Matthews)*

Slimbridge

messages in the ancient civilizations of the Mediterranean and the Middle East and in more recent centuries they have been bred for their ability to home rapidly in competitive pigeon racing. Their remarkable ability to home hundreds of miles from an unfamiliar release site to their home loft raises the question of what mechanism they are using. We shall consider the main hypotheses in the rest of this chapter.

As we consider the evidence, we should keep in mind that much of it is controversial. Little is agreed on, from fundamental data to the evidence for particular hypothetical mechanisms. Robin Baker in 1984 challenged whether the observations on pigeons require any true navigating skill at all. He suggested that the "unfamiliar" release sites of pigeons may in fact be familiar, and that the pigeons home using remembered landmarks and home cues (whether visual, auditory, olfactory, or magnetic). He claimed that none of the release sites were sufficiently far from the home loft to rule out the use of such cues, except in a few cases, such as trans-Atlantic displacement, and in these the evidence of homing was very weak. Apart from landmarks, pigeons could also home without any great navigational skill if they learned the direction on the way out. But this cannot be the only explanation, because pigeons can home normally even when taken out in enclosed vans in continuously rotating cages, or even under anesthetic; both treatments should render learning the outward journey impossible.

Most workers in the field, however, assume that pigeons can home from unfamiliar starting points, and we shall accept that assumption (at least for the sake of argument) from here on. Human beings who tried to navigate home from an unfamiliar site would need a map and a compass. They would then remember the map location of home, identify their present map position, and use the compass to find the direction from one to the other. Experimenters on pigeons have accordingly searched for a "map sense" and a "compass sense." The compass sense is the less controversial. Many experiments suggest that pigeons, as their first resort, use the position of the sun in the sky, making allowance for its daily changes by means of an internal clock. Thus, if a pigeon is trained on an artificial light/dark schedule different from the outside, it will navigate with predictable error beneath the natural sun: pigeons, for instance, whose internal clock is shifted six hours forward will head off 90° counterclock-wise. away from the correct direction (Figure 5.8). Pigeons used to the northern hemisphere but moved to the southern hemisphere likewise orient with the predicted 180° error at noon. But pigeons must possess some other compass sense as well as the sun, because they can navigate correctly on overcast days, when the position of the sun is invisible. (We shall meet some evidence that a magnetic sense is at least one such back-up, in Figure 5.9.) The experiments with clock-shifted pigeons, however, strongly suggest that the sun supplies their preferred compass.

It is the map sense that is the real mystery. Many imaginative possibili-

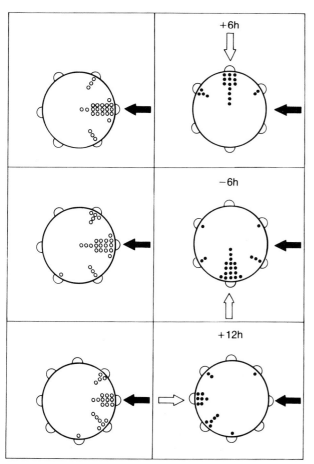

Figure 5.8
Clock shifts alter the orientation of homing pigeons. In this experiment, pigeons have been trained to peck toward a particular compass direction, which is that of the controls illustrated to the left. Each dot shows one peck by one pigeon. The directions of pecking by clock-shifted birds are shown on the right. The directions are those that would be expected if the pigeons use the sun as a compass, making allowance for the movements of the sun through the day.
(After Baker)

ties have been investigated, but we shall concentrate on three here: the sun, the earth's magnetic field, and olfactory home cues. Pigeons may well make use of a number of sensory mechanisms, perhaps including all three considered here and some others too; and they may be able to exploit any combination of the mechanisms, depending on the best cues at their location at the time they are flying. The different mechanisms, therefore, are not simply alternatives. At present the evidence counts against the "sun arc" hypothesis, but recent experiments increasingly suggest that homing pigeons can make use of magnetic and olfactory cues.

Let us take the "sun arc" hypothesis first. We have seen that the position of the sun is used as a compass by which pigeons direct themselves away from release points. It could also be used as a map. A map is most easily conceived as having two coordinates, like longitude and latitude on customary human maps. The sun arc hypothesis of E.V.T. Matthews suggests how the position of the sun could supply longitudinal and latitudinal positions. The pigeon would have to remember the position of the sun above its home loft at each time of the day. Then, at a release site, it could estimate the latitude by the height of the sun in the sky (through an estimated arc) and longitude by the time of day indicated by the sun at the release site, relative to the time of day at home as indicated by the pigeon's internal clock. If its home clock said the time was 12:00 A.M. but the sun at the release site indicated 6:00 A.M., the pigeon would infer it had been moved one-quarter of the way around the world to the west. The hypothesis can therefore be tested by clock-shifting experiments.

Do clock-shifted pigeons misorient in the way the sun arc hypothesis predicts? A pigeon that has been clock-shifted six hours forward should behave as if it had been taken west: a pigeon, for example, clock-shifted six hours in a loft at Rome and then released nearby outside the loft at noon would think that the time at home was 6:00 P.M. It would deduce that it had been moved six hours west (to the east coast of the United States), and therefore would home by flying east. (We ignore the fact that in this simple experiment it would recognize local landmarks.) Even if displaced through three time zones eastward, a pigeon would still infer it was over the Atlantic, and home in the wrong direction. However, when the appropriate experiment was performed, the hypothesis proved wrong. Only the pigeon's compass sense was disrupted, as in the previous clock-shifting experiments we discussed: its map sense was unaffected. Thus, pigeons clock-shifted six hours early and taken three time zones to the east behaved as if they "knew" they should home to the west. Their compass bearings were rotated 90° clockwise because their internal clocks pointed to midday when it was really 6:00 P.M., and they actually flew north rather than west. The relevant experiment, by Walcott and Michener, has not escaped criticism, but it is not the only piece of evidence to go against the sun arc hypothesis; even if that hypothesis has not been definitely refuted, we can say that such evidence as there is counts against it.

Magnetic polarity can clearly provide one coordinate of a map sense; various possibilities have been offered as to how magnetism might provide a second coordinate to complete the map. We shall not discuss the physical details here, only the evidence on the magnetic sense of pigeons. It is certainly true that pigeons with bar magnets attached to their heads navigate incorrectly on overcast, but not on sunny, days (Figure 5.9); this suggests both that pigeons have a magnetic sense, and that it may be used by pigeons as a back-up compass sense when the sun is invisible.

Figure 5.9

The orientation of homing pigeons is influenced by changes in the magnetic field. The dots represent the angles at which pigeons flew off from the release site: the direction home is straight up; straight down represents 180° away from home. There are two magnetic conditions, which do not affect orientation on sunny days (left) but do on overcast days (right). *(After Baker)*

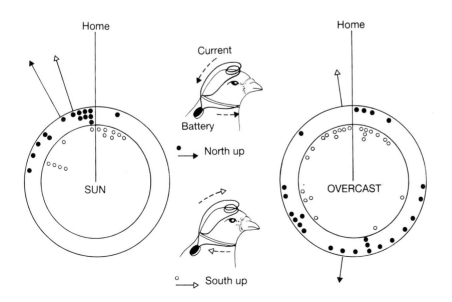

Moreover, pigeons navigate less accurately when the earth's magnetic field is disturbed (for instance, by solar flares), but this evidence is inconclusive, because other factors may be correlated with these natural disturbances. Whether pigeons use a magnetic compass is still a controversial topic, but a magnetic map sense is altogether more speculative; at present it is a theoretical suggestion. It does raise the question of how pigeons detect the magnetic field. No magnetic sense organ has been identified, but two hypotheses have been put forward. One is that the light-sensitive pigments of the eye could also act as magnetic sense organs. The other is that the molecule magnetite may act as the magnetic sense, and indeed traces of magnetite have been found in bones and brains of pigeons and other species, including humans. This research is at an exciting stage, and the people carrying it out hope to identify magnetic sense organs in the next few years. Meanwhile, the magnetic hypothesis of homing has yet to be confirmed.

Finally, there is the possibility that homing pigeons use olfactory cues. According to this hypothesis, pigeons learn a home odor, or a "landscape" of orders around the home loft, and orient with respect to them. Pigeon homing could then be a case of a learned home cue, like salmon homing. Alternatively, pigeons might have a sort of olfactory map, which they could use to navigate. A first suggestive result is that pigeons home less accurately if their olfactory sense is experimentally removed (or strongly reduced). The design of such an experiment is to cut one of the olfactory nerves—there are two nerves, one per nostril—and block up the opposite nostril; the pigeon can then breathe only through the nostril on the side its olfactory nerve has been severed. Control birds have one olfactory nerve cut, but on the same side as their nostril is blocked: they can smell through the nostril they breathe through. The control birds home almost as well as normal pigeons, but the experimental birds home much less efficiently. Another suggestive result is that the homing ability of pigeons can be reduced by distracting them with a strong smell at the release site. Figure 5.10 illustrates another experiment, in which pigeons had the aromatic chemical benzaldehyde blown at their home cages from a north-northwest direction. In the illustrated experiment, the pigeons were then taken 50 km east and released in an

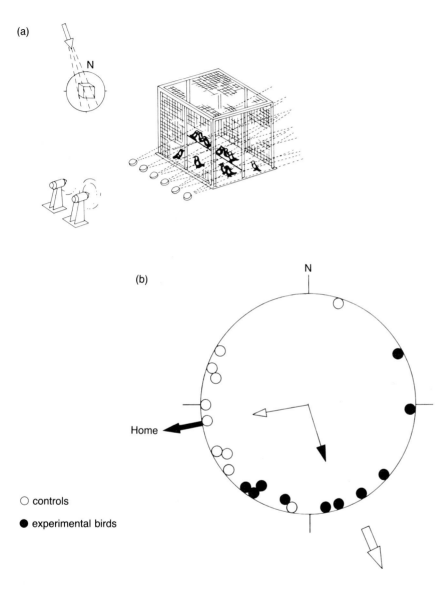

(a)

N

(b)

N

Home

○ controls

● experimental birds

Figure 5.10
Evidence for use of olfaction in pigeon homing. (a) Experimental apparatus: pigeons have benzaldehyde blown at them from a north-northwest direction. (b) When taken east to a release site, the pigeons had benzaldehyde painted on their beaks. Experimental pigeons that had benzaldehyde blown at them (as in (a)) homed south-southeast; control birds that had not had benzaldehyde blown at them at home homed correctly to the west. The arrows inside the circle indicate the average direction of homing. The arrows outside the circle indicate the true direction of homing (black arrow) and the direction opposite to that from which the experimental pigeons had benzaldehyde vapor blown at them in their home loft (open arrow). (*After Ioalè, Nozzolini, and Papi*)

unfamiliar place; at the release site they had benzaldehyde painted on their beaks. The pigeons apparently deduced that they must be north-northwest of home (at the source of the smell they were used to) and flew off south-southeast, rather than west. Control birds, who had not had benzaldehyde blown at them while in their lofts, but did have it painted on their beaks at the release site, homed normally.

Despite these striking results, students of the subject have been slow to draw firm conclusions. There has been a historical tendency for experiments not to be confirmed when repeated elsewhere, or to yield erratic results, and single experiments are therefore unconvincing. Moreover, in the case of the olfactory hypothesis, it has proved impossible to train pigeons to make the relevant olfactory distinctions (which is a standard, and powerful, method of testing for sensory abilities—p. 67–68). We can conclude, therefore, that olfactory cues probably are used in some circumstances, but how important they are in general is undecided.

The most plausible explanation for conflicting results (and the results we have looked at here are only a minor part of the huge number of experiments that have been done) is that homing pigeons can use a large number of cues and the particular cues they use in any one case depend on the local conditions. Olfactory cues may be important in an experiment in which pigeons are exposed to strong smells, but not (perhaps) if the smells are weaker or the information from other sources is more clear-cut. Pigeons undoubtedly possess an excellent ability to home, and they make use of landmarks and other home cues when they can be sensed. They have a compass sense, primarily from the sun, but they also have a back-up compass sense when the sun is invisible. Their apparent ability to navigate in the full sense of being able to home from unfamiliar starting points is probably based on a variety of cues, including the sun, magnetism, and olfaction. Questions remain open about how these cues are used (and indeed, whether they are used at all), and of what the pigeons' priorities are when faced with a variety of conflicting sensory information.

5.4   Summary

1. Animals migrate in order to find environments superior to the one they are in. When environments change, migration may become advantageous.

2. Migrating animals need a mechanism to locate superior environments. Wildebeest rely on sensing distant rainfall, which precedes environmental improvement. Seasonal changes are anticipated by birds through the changes of day length.

3. Emigration from overcrowded areas can result in the regulation of population densities.
4. Bee-killing digger wasps find their way home by memorizing the landmarks around their burrow entrance. Salmon probably find theirs by the memorized odor of their riverine birthplace.
5. The most versatile homefinders are pigeons, which may be able to find their way home from unfamiliar starting points. They seem to be able to navigate home even when taken to the release point under such conditions as would make learning the way out impossible.
6. Pigeons probably navigate home by means of "compass" and "map," although the need for a map sense has been challenged. They use the sun (and perhaps the earth's magnetic field when the sun is invisible) as a compass. Their map sense is unknown: the sun, the earth's magnetic fields, and olfactory maps have all been suggested and experimented on.

5.5   Further reading

Waterman (1989) discusses most of the topics in this chapter in more depth, as does Baker (1982, 1984). Berthold (1991) and Papi (1992) provide modern reviews of research on migration; Berthold, in particular, has done interesting selection experiments on migration in blackcaps, and his results illustrate an extra theme the chapter did not have space for—the genetics of migration. Tinbergen (1958) describes his delightful experiments on *Philanthus*, only one of which did I have space to explain. Stabell (1984) and Thorpe (1988) review salmon homing. Lohmann (1992) discusses the further question of navigation in sea turtles. Guilford (1993) is a short overview of modern work on magnetic sense and on pigeon homing.

CHAPTER 6

Eating and Not Being

Eaten

Many of the observable behavior patterns of animals are concerned with feeding, or avoiding being fed on. We begin by looking at a number of adaptations for finding, recognizing, and catching prey, and then turn to antipredatory behavior. The feeding adaptations of particular species depend on the nature of their ecology.

6.1   Eating

The abstract principle of feeding is simple: to find and catch food for yourself, while not being caught as food by another animal. The exact techniques used by different species are determined by the nature of their diets and predators—that is, by their ecology (ecology is the science that deals with the relations of living species, and communities of species, with their environments). To understand the diversity of behavioral adaptations for feeding, therefore, we must understand the ecological divisions of feeding types. A first distinction is between herbivores and carnivores. The energy input to the world comes from the sun: plants utilize the sun's energy by photosynthesis, herbivores feed directly on the plants, and carnivores on the herbivores. The chain has further steps. Carnivores feed on other carnivores (which make up 10% of the diet of the leopard, for example), scavengers feed on the dead carcasses of all kinds of animals, and decomposing fungi and bacteria return the nutrients from dead bodies to the soil—where they can be reused by plants.

The behavioral adaptations of herbivores and carnivores will differ. The herbivore has no difficulty in catching food, though it may be difficult to select the edible parts of plants, as they are frequently tough, indigestible, short of nutritive value, and stuffed with poisons. The carnivore at least has to be mobile to catch its prey. However, let us turn to a more detailed example, to see how a species' feeding behavior evolves as part of its general feeding ecology.

6.2　Feeding in group-living herbivores

Buffalo, zebra, wildebeest, topi, and Thomson's gazelle live in huge groups that together make up some 90% of the total weight of mammals living on the Serengeti Plains of East Africa. They are all herbivores, and they all appear to be living on the same species of grasses, herbs, and small bushes. The appearance, however, is illusory. When Bell and his colleagues analyzed the stomach contents of four of the five species (they did not study buffalo), they found that each species was living on a different part of the vegetation. The different vegetational parts differ in their food qualities: lower down, there are succulent, nutritious leaves; higher up are the harder stems. There are also sparsely distributed, highly nutritious fruits, and Bell found that only the Thomson's gazelle eat much of these. The other three species differ in the proportions of lower leaves and higher stems that they eat: zebra eat the most stem matter, wildebeest eat the most leaves, and topi are intermediate.

How are we to understand their different feeding preferences? The answer lies in two associated differences among the species, in their digestive systems and body sizes. The digestive systems can be divided into the nonruminants (the zebra, which has a digestive system like a horse) and the ruminants (wildebeest, topi, and gazelle, which are like the cow). Nonruminants cannot extract much energy from the hard parts of a plant; however, this is more than made up for by the fast speed at which food passes through their guts. Thus, when there is only a short supply of poor-quality food, the wildebeest, topi, and gazelle enjoy an advantage. They are ruminants and have a special structure in their stomachs (the rumen), which contains microorganisms that can break down the hard parts of plants. Food passes only slowly through the ruminant's gut because ruminating—digesting the hard parts—takes time. The ruminant continually regurgitates food from its stomach back to its mouth to chew it up further (that is what a cow is doing when "chewing cud"). Only when it has been chewed up and digested almost to a liquid can the food pass through the rumen, and on through the gut. Larger particles cannot pass through until they have been chewed down to size. Therefore, when food is in short supply, a ruminant can last longer than a nonruminant because it can extract more energy out of the same food. The differences partially can explain the eating habits of the Serengeti herbivores. The zebra chooses areas where there is more low-quality food. It migrates first to unexploited areas and chomps the abundant low-quality stems before moving on. It is a fast-in/fast-out feeder, relying on a high throughput of incompletely digested food. By the time the wildebeest (and other ruminants) arrive, the grazing and trampling of the zebras will have worn the vegetation down. As the ruminants then set to work they eat down to the lower, leafier parts of the vegetation. All of this fits in with the differences of stomach contents with which we began.

The other part of the explanation is body size. Larger animals require more food than smaller animals, but smaller animals have a higher metabolic rate. Smaller animals can therefore live where there is less food, provided that it is of high energy content. That is why the smallest of the herbivores, Thomson's gazelle, lives on fruit, which is very nutritious but too thin on the ground to support a larger animal. By contrast, the large zebra lives on the masses of low-quality stem material.

The differences in feeding preferences lead, in turn, to differences in migratory habits. We have seen (p. 120) that wildebeest follow, in their migration, the pattern of local rainfall. The other species do likewise. But when a new area is fueled by rain, the mammals migrate toward it in an orderly pattern to exploit it. The larger, less fastidious feeders, the zebras, move in first; the choosier, smaller wildebeest come later; and the smallest species of all, Thomson's gazelle, arrives last (Figure 6.1). The later species depend on the preparations of the earlier, for the actions of the zebra alters the vegetation to suit the stomachs of the wildebeest and gazelle.

If we are to understand the feeding habits of the species, therefore, we must consider it in relation to the whole ecology of the species, and its relations to other species. Behavior is an inseparable part of a whole system, made up (in this example) of body size, gut morphology, and the habits of associated species.

6.3   Recognizing food

Given that the diet of a species is determined by its ecology, the individual members have two behavioral problems, the best solutions to which will depend on their diet. Food must first be recognized, and then caught. We shall consider these two in the next two sections. First, how is food to be found? The environment of an animal contains all sorts of things, some of which are edible and some of which are not. The problem for the hungry animal is to distinguish the former from the latter, and to put only the edible things in its mouth. What is food for one species is often poisonous, or indigestible, for another. Each species must recognize its own kind of food. The complexity of the problem depends among other things on whether the species is a specialist feeder, which eats only one or two types of prey, or whether it is an omnivore, which eats many kinds of prey.

Let us consider first a fairly specialist feeder, the toad. Toads mainly eat small, dark arthropods, such as millipedes and beetles, and earthworms. They stick out their tongues and snap at any such prey that passes by. J.-P. Ewert and his colleagues have studied in detail how the toad recognizes its meal. The toad does not recognize the prey as such, but recognizes small, dark, moving objects. If a piece of dark paper about ½ inch in length is moved where a toad can see it, it will snap at it just as if it were prey (Figure 6.2). This recognition system works for a toad because in nature small, dark, moving objects *are* prey. Only in the lab is it fooled by bits of paper. Ewert and his colleagues have also worked out where, in its nervous system, the toad recognizes small,

Figure 6.1

The main species of large herbivore of the Serengeti Plains, East Africa, migrate into an area after rainfall in a predictable order: after the rain, the grass grows; zebra then arrive first, followed by wildebeest, and Thomson's gazelle. The order of migration can be explained by the different diets of the different species.
(After Bell)

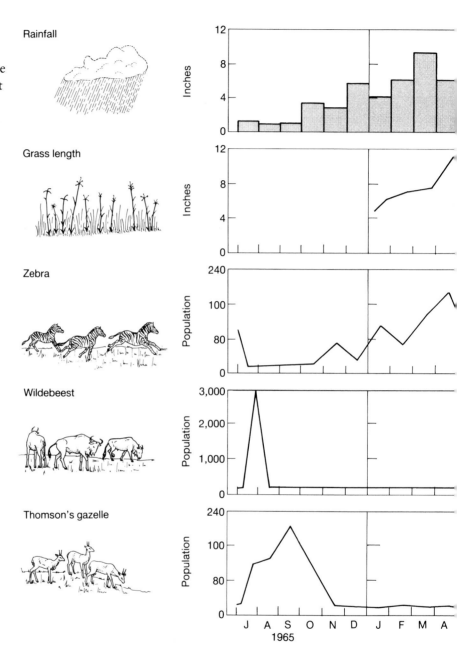

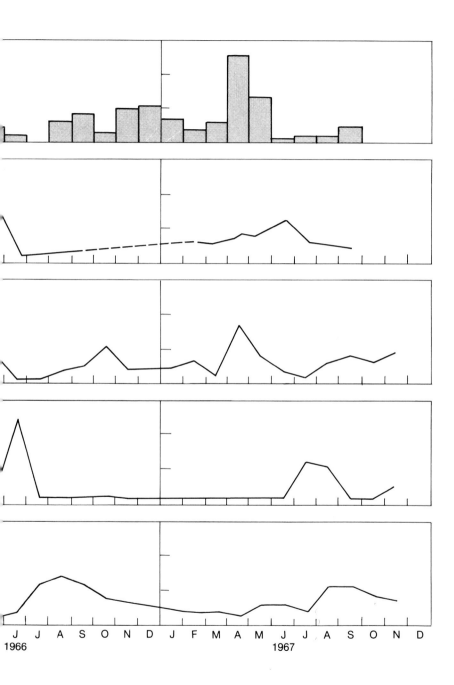

J J A S O N D J F M A M J J A S O N D
1966   1967

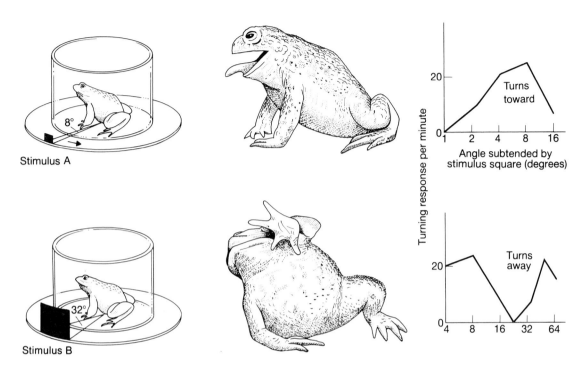

Figure 6.2   Toads recognize small moving objects as food. In the experiment illustrated at the top, the toad's response to a small black object revolving in its visual field is measured. The graph at the right reveals that the response depends on the apparent size of the object. If it is apparently small, the toad orients toward it; if it is apparently large, the toad avoids it. *(After Ewert)*

dark, moving objects. They found neurons in the retina of the eye that respond differently according to the size of dark objects moving in the field of view. There are three classes of sensory neurons that respond to three different classes of objects. The visual system of a toad is a world of moving dark objects of various sizes and shapes; its gastronomic choice is to put all the dark objects of a certain size in its mouth.

The toad's response to different-sized objects is not much affected by learning; to find cases where learning has an influence we must turn to animals with more catholic diets. The first learning phenomenon to consider is a perceptual one, called a search image. Forming a search image means learning to see something that had not previously been seen. It often occurs in humans with photographs of camouflaged animals. To begin with you cannot see the animal at all, but after you have first noticed it, it becomes obvious. You have learned to see it. It is more difficult to know whether an animal goes through the experience of at one moment not being able to see a food item, but then being able to see it at the next. A human can tell you of this experience verbally; with animals you can only study what they do. If an animal does not eat a food item at one moment but does later, it may only be because it was not hungry before, not because it has learned to see it. An experiment by Marian Dawkins probably demonstrates the formation of a search image. She worked with domestic chicks feeding on rice grains. She dyed some of the rice grains a different color from the background, so that these grains were easy to see; she dyed other grains the same color as the background, which made them difficult to see. The chicks pecked the conspicuous rice grains, which proves that they were eager to eat rice. However, there was a delay of a few minutes from the start of the experiment before the chicks ate their first camouflaged rice grains. After eating one camouflaged grain, they rapidly ate more of them. To be precise, there was an average delay of 66 seconds from eating a visible grain to eating a camouflaged one, but a delay of only 6.7 seconds from eating one camouflaged grain to eating another camouflaged one. The chicks learned to see camouflaged grains: they formed a search image for them.

Learning becomes even more important in the feeding of so omnivorous an animal as the rat. Rats will initially sample, in small quantities, almost anything, but if they find what they have taken makes them sick, they avoid eating anything like it again. This "one-trial" learning, however, does not operate in the perceptual system of the rat, but in its decision-making machinery. The simple recognition system of the toad clearly would be utterly inadequate for an animal that must be able to distinguish between a large number of potentially edible objects.

6.4 Hunting food

6.4.1 Foraging on immobile prey

Animals that feed on static food have to move around to find it, and, as we shall see, the best course of movement depends on the distribution of the food items in space. If the prey can themselves move, the predator will have to be able to move faster, or with sufficient stealth that it can catch its prey without a long chase. It may pay to hunt in groups; prey can then be surrounded rather than chased, and large species of prey can be subdued by a number of smaller predators. Other animals, such as spiders, caddis flies, mantises, and angler fish, do not move to catch their mobile prey. They "sit and wait," and ambush prey that move to them.

There is a general principle of feeding that the animal's pattern of movement—hunting, foraging, or searching—should fit the distribution of prey. This is now a flourishing area of research, but we shall pick only one illustration. J.N.M. Smith put out pastry "caterpillars" for thrushes and watched how the birds adjusted their movements to the arrangement of prey. The prey could be placed with regular spacing, at random, or clumped in groups. Smith found that the thrushes moved differently when the prey were clumped than when they were spaced out (Figure 6.3). Thrushes move in discrete jumps, of which Smith could measure two properties: their length, and the angle of turn between two jumps. When food was clumped, a thrush, after finding a food item, turned

Figure 6.3　　Area-restricted search. Birds can adjust their movements according to the distribution of food. At the left the movements of a ground-feeding bird, such as a song thrush (*Turdus philomelos*), are shown for a bird that has learned that food is clumped. It strides more or less straight ahead until it finds a food item, and then changes its movements, now turning through tight angles. The movements illustrated on the right are those of a bird that has learned its food is spaced out. It keeps walking straight ahead even after it has found a food item.

through a sharper angle than when it was spaced out, but kept the length of each jump constant, independent of the food distribution. The turning angles used by the thrush make sense. When food is clumped in groups, the best method of search is to move straight ahead when not in a group, but to turn through tight angles once one item has been found, in order to find the rest of the group. However, when food is spaced out, the bird is unlikely to find another food item nearby after finding one, and it is best to keep moving on. The change in behavior for clumped food items is called area-restricted searching. It is an adaptation to increase the rate of finding food.

6.4.2 Group-hunting carnivores

By no means all carnivores that live on mobile prey live in groups. But we do not have space to consider all feeding methods and will pick group hunters as one example of the active carnivorous habit (Figure 6.4). It will incidentally allow us to consider the question of what advantage group living has. A pride of lions hunting down a prey animal, such as a zebra, is one of nature's more awesome spectacles. A complete pride might contain two males, seven females, and some cubs, but the hunt is female business. The males stay behind, for their showy manes would only disturb the hunt, whose technique is stealth.

Figure 6.4   A pride of lionesses on the Serengeti Plains, Kenya. On large prey, lionesses hunt more successfully in coordinated groups. *(Photo: Heather Angel)*

The hunt starts with the lionesses hunting as a group for their prey. When they spot some promising zebras or antelopes, they spread out into a line. They now move slowly, stealthily toward their unknowing victims. Soon one lioness is close enough to a zebra to attack. In the ensuing panic zebras run in all directions, many of them right into the paws of another waiting lioness. The lionesses also cooperate in killing prey. If the prey is large and dangerous, such as a buffalo, the attack of several lionesses is the safest method of execution.

Not all group hunts are successful. In fact, of the group hunts watched by Schaller on the Serengeti Plains, only about 30% were successful. Lionesses hunting in groups are, however, more successful than lionesses hunting alone. Lionesses often hunt alone, but only about 15% of these hunts are successful. The advantage to carnivores such as lions of hunting in groups is that they can catch more food than if they hunted alone. They can also catch kinds of prey that they could not catch by themselves, such as buffalo. Carnivores can also adjust the size of the hunting party to the kind of prey being sought. Hyenas, to take another example, hunt in groups for zebra, but hunt singly for the smaller Thomson's gazelle.

Lions are strong but cannot run for a long time. A zebra or antelope could easily outrun a lion, and lions must therefore rely on stealth and the surprise attack; carnivorous dogs, by contrast, hunt by running down their prey. The African hunting dog and the wolf employ similar hunting techniques. Wolves hunting moose, or dogs hunting wildebeest, start by charging several of the prey in a group. The prey easily run away, but the dogs have a chance to select a vulnerable individual. They single out an old, young, or infirm animal and only then start the chase. During the chase, the dogs of the pack may take turns in leading. When the kill comes they usually all join in to wear their victim down. They are flexible, however, and employ other techniques when appropriate. They pay particular attention to prey that straggle or leave the herd. One dog may then try to get between the herd and the straggler, and try to drive it toward the rest of the pack.

6.4.3 A note on domestication

Domestic dogs are descendants of wolves, to which they show many similarities of appearance and behavior. Their social habits have fitted

them to human society. The exact reasons why dogs were first domesticated by humans are now lost in the past, whether it was the agreeable
companionship of an animal that apparently respects its master much as a
dominant member of its hunting pack, the exceptional sensitivity of their
whole body surface, their uses in agriculture, or, most likely, a combination of all such factors. Undoubtedly their ancestry rendered dogs particularly good pets. We still exploit the ancestral habits of dogs in their training. The hunting techniques we have just discussed for wolves and African
dogs are made use of in the training of sheepdogs. For instance, young and
untrained sheepdogs will often spontaneously run around to the other
side of a flock of sheep and try to drive them toward the shepherd. It is
only a small step from this to teach the sheepdog to "circle," to run around
the flock and keep the sheep tightly bunched together (Figure 6.5).

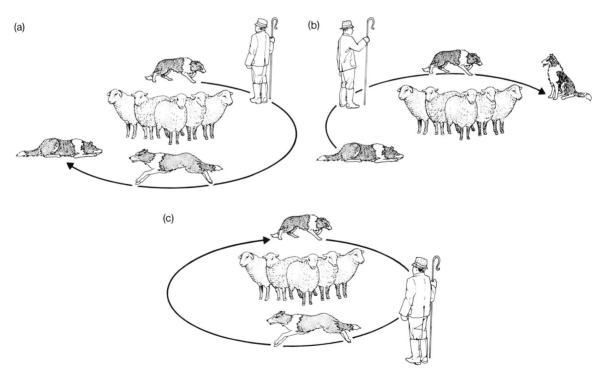

Figure 6.5   Learning of "circling" in sheepdogs. A naive dog will spontaneously run around
to the other side of the flock. It can therefore be taught to circle if (a) the shepherd first makes the
dog run around to the other side of the flock, and then (b) moves himself and repeats the exercise.
Soon (c) the dog will encircle the flock by himself. *(After Vines)*

Sheepdogs are also particularly alert to stray sheep and can easily be taught to drive stray sheep back to the shepherd. If we imagine that the sheepdog is treating the shepherd as a fellow hunter, then circling and retrieving stray sheep both manifestly resemble the ancestral canine hunting behavior. Moreover, the most difficult tricks to teach a sheepdog are those most removed from its ancestral hunting skills. Sheepdogs are difficult to teach to drive the flock away from the shepherd—the dog has to be restrained from its desire to circle around and drive them back. Sheepdogs also are difficult to teach to leave stray sheep they have gathered, so they can go and gather more.

The learning abilities of the sheepdog are closely related to its ancestry. It learns most easily what comes naturally to it. This is a general principle in teaching tricks to animals. Circus animals, for another example, learn tricks most easily if those tricks are a simple extension of the animals' natural behavior.

6.5   Avoiding being eaten: active evasion

Any property of an organism that reduces its chances of being taken by a predator will be favored by natural selection. The resulting antipredator adaptations are very diverse. In the following four sections we shall examine five different habits used by animals to avoid being eaten: potential prey may actively flee their predators; they may stay still and try to be invisible; or they may stuff themselves with sickening chemicals and advertise their unpalatability with bright "warning colors," or they may mimic the warning colors of others; and finally, in some circumstances, an animal may make itself less likely to be eaten by living in a group.

Active flight is used by many animals to escape predators, and a particularly elegant study has been made on noctuid moths by Roeder. Noctuid moths are eaten by bats and have evolved a special pair of ears to warn them of approaching danger. There is one ear on each side of the thorax and each has a simple structure; two nerves connect each ear with the thoracic ganglion (which is the nearest minibrain). They are sensitive to the high-pitched squeaks used by bats to find their prey, and they have an advantage over the bat, in that the bat emits very loud blasts in order to detect a faint echo. The moth can hear the bat from a greater distance than the bat can pick up the echo from a flying moth; to be precise, a moth can hear a bat about 100 feet away, whereas a bat can detect a moth at a range of less than 8 feet. The moth, moreover, can tell whether the bat is to the right or left (because it has an ear on each side) and whether it is approaching or moving away. A bat approaching a moth will sound louder and louder as it comes close, and the moth is sensitive to loudness. Bats do not fly in the same direction for long. Therefore, if a moth hears a bat approaching about 100 feet away, its best policy is to fly off in the other direction. That way it may get out of the bat's flight path before it enters the detection range. Once a bat has detected a moth it has the advantage, because bats can fly much faster than moths. The moth's surest means of staying alive, therefore, is not to be detected, and it has the advantage of advance warning to keep out of the way.

A bat may appear suddenly out of the dark close to a moth. It is then useless for the moth to flee, because it will have been detected and the bat can fly faster than it. When a moth detects a loud bat sound, indicating a bat less than about 8 feet away, it puts a different escape tactic into action. It flies in wild loops and spirals, and dives to the ground, a course of flight designed to make it as difficult as possible for the bat to catch it. (The erratic flight may be produced by the moth by just switching off its steering mechanism. Then even the moth will not know where it is going: the most effective means of confusing someone else about where you are going is not to know yourself.) The moth has two escape responses: if it hears a bat afar, it turns and flees; if it is surprised by one nearby, it goes into a crazy flight. It uses its hearing sense to decide which response is appropriate.

6.6 Camouflage

The noctuid moth's defense is to seek escape in active flight. The opposite defense is to sit dead still and try to be invisible. Such is the method of camouflage in which a species evolves to resemble its background.

Camouflage is of course an adaptation of appearance and coloration, but the most exquisite artistry will be wasted if the animal's behavior is not suited to the camouflage. The world is a patchwork of different colors: the animal is only camouflaged if it settles in the right place. Consider the European grasshopper *Acrida turrita*. It comes in a green form and a yellow form. In nature the green form lives in green places and the yellow form in yellow and brown places, with rare exceptions. In a simple experiment, the German ethologist S. Ergene gave yellow and green grasshoppers a choice between yellow and green backgrounds. The green grasshoppers fittingly tended to settle on the green backgrounds, and the yellow grasshoppers on the yellow.

The North American moth *Melanolophia canadaria* faces a more difficult problem in lining up with its background. It has striped wings and lives on the bark of trees. It must line its stripes up with the lines of the bark if it is to be camouflaged. In an experiment, T.D. Sargent allowed the moths to sit on cylinders that had regions of vertical stripes and regions of horizontal stripes. If the stripes (which were made of black tape stuck on a white surface) could be felt by the moths, then the moths usually lined up correctly. When Sargent covered up the stripes and surface with a transparent film, the moths no longer lined up correctly. The moths must be relying on the feel of the surface that they have to line up on. In nature they will be able to feel the stripes of their background, and ensure that they settle in a camouflaged posture.

6.7   Warning coloration and mimicry

Some animals protect themselves against being eaten by containing poisonous or sickening substances. Some such animals make their own poisons; others take them from sources in their environment. The wings of the monarch butterfly, for example, contain powerful heart-stopping poisons called cardiac glycosides. The monarch eats the poisons as a caterpillar, when its food plant is the asclepiad, or "milkweed," which contains cardiac glycosides. The caterpillar is not harmed by the poisons; it just stores them, and they are then retained by the adult. The monarch is also brightly colored, both as a caterpillar and adult. This bright coloration poses several evolutionary problems, and we can look at the two main ones. The first is what advantage is gained by bright coloration. The answer is probably that predators learn more quickly to avoid poisonous prey if they are brightly colored. When a bird eats a monarch, as an adult or caterpillar, it will be violently sick within minutes, an experience it would learn not to repeat. The bird's problem then is to distinguish sickening from edible prey. Now, the colors of the monarch are bright gold, and it may be that the bright coloration evolved to make the monarch more memorable to birds. It is therefore called "warning coloration." Experiments have shown that predators learn to avoid sickening prey. Jane Brower, for instance, offered the monarch butterfly as food to the Florida scrub jays (*Aphelocoma coerulescens*). On their first meal of monarchs the jays were violently sick, but after only a few trials they had learned not to eat monarchs, though they continued to eat other, tasty food.

However, Brower's experiment showed only that the birds learned to avoid eating monarchs: it does not show that conspicuous coloration is any more effective than dull coloration in teaching birds what to avoid. To show that, we need an experiment in which both conspicuous and inconspicuous (or cryptic) food is given to a predator, and both kinds are offered in distasteful and edible forms. Then we can see whether the predator learns to avoid the conspicuous distasteful prey more effectively than the cryptic distasteful prey. T.J. Roper and S. Redston did the necessary experiment. They used domestic chicks, and their experiment had two stages, a "training" stage and a "test" stage. In the training stage, a chick was allowed to peck once at a bead that was either red (and cryptic) or white (and conspicuous) on a red background. (Chicks will peck at a bead as if it were food.) Each color of bead in turn could have been coated either with a distasteful substance (methyl anthranilate) or left untreated: these are called the "experimental" and "control" treatments. Then, a little while later, in the test stage, the chick was left for four minutes with beads of one of the same color as it had been trained with, and the rate of pecking was measured. The rate of pecking was much lower in the experimental case with conspicuous beads than in either controls or in the experiment with cryptic beads (Figure 6.6). We

Figure 6.6

Conspicuousness enhances learning in chicks. Chicks that have previously experienced a distasteful bead (experiment) peck at beads at a lower rate than controls (in which the beads were untreated) and than chicks that have experienced cryptic distasteful beads. *(After Roper and Redston)*

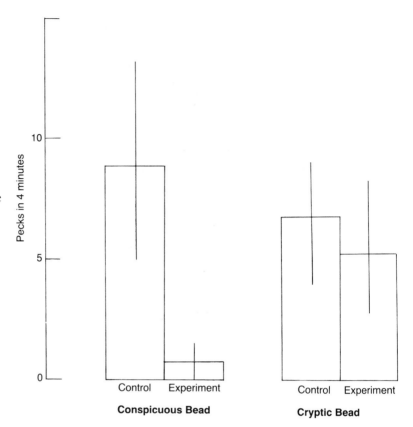

can conclude that the conspicuousness enhanced the learning by the chick.

Roper and Redston's experiment was designed to test not only whether conspicuousness is advantageous in distasteful prey, but also whether the advantage is specifically through its effect on the predator's learning mechanism. It could be that distasteful conspicuous prey are avoided more simply because the predator sees them more easily and therefore eats more of them early on, and thus has a better chance to learn about their distasteful qualities. However, Roper and Redston allowed their chicks to peck only once in the training stage, and therefore all four kinds of chicks had quantitatively the same experience with their type of bead. Thus, the decreased pecking rate shows specifically that the chicks learn to associate conspicuousness and distastefulness more readily than they learn to associate cryptic color and distastefulness. It is as if the conspicuous, disgusting prey sticks more powerfully in their memory.

The early stages in the evolution of warning coloration pose another problem. A population of monarchs clearly benefits from teaching each generation of birds not to eat them. They benefit by being eaten less. The problem is that some butterflies have to be eaten to teach the birds to begin with. In evolutionary terms, it is no consolation to the dead butterflies that some others are benefiting from their death. They are dead; they have failed to reproduce, and natural selection has worked against them. When warning coloration first appeared in evolution, it would have been a rare, minority trait. Those rare, brightly colored butterflies would have been conspicuous to predators, and therefore eaten. One would expect natural selection to have eliminated the trait. How could natural selection favor an increase in its frequency? The question has not been satisfactorily answered, though there are some possible explanations. One possible answer is that the family relatives of the eaten butterfly may benefit from its death. Members of the same family tend to resemble each other; if one were warningly colored, many others would be as well. If one member were killed, the rest of the family would benefit from the lesson taught to the predator. (This process would be an instance of "kin selection": see Chapter 10.)

However that may be, the learning by predators to avoid prey of a particular appearance makes possible the evolution of another strategy: mimicry. Predators, we have seen, learn not to eat monarchs because they are sickening. But another butterfly, the viceroy, inhabits some parts of the monarch's geographic range, and is not poisonous to birds. It also looks very much like a monarch. It can presumably survive, even though it is edible to birds, because birds take it for a monarch and avoid it. An experiment by Brower suggests this is so. She divided eight Florida scrub jays into two groups of four each. She first offered monarchs to four experimental birds and viceroys to four control birds; she recorded how many times the birds avoided the butterflies or pecked at them. The experimental jays should have learned that the color pattern is associated with unpalatable food; the control jays should not. Brower then offered viceroys to both classes of jays and recorded whether they

---

Table 6.1 Experimental demonstration of mimicry. Brower had offered four Florida scrub jays ("control" birds) palatable viceroy butterflies and four other Florida scrub jays ("experimental" birds) sickening monarch butterflies. The monarch and viceroy look alike. Brower then offered viceroys to both classes of jays and recorded the number of trials in which they avoided the butterflies or pecked at them; the numbers in the table give her results. The difference is statistically significant. (*Slightly simplified from Brower, 1958.*)

	Control birds				*Experimental birds*			
Number of trials in which the jay:	C-1	C-2	C-3	C-4	E-1	E-2	E-3	E-4
Avoided viceroys	0	0	9	1	14	12	12	4
Pecked viceroys	25	25	16	24	2	1	3	12

avoided or pecked at them. The experimental birds now tended to avoid the viceroys, and the control birds to peck at them (Table 6.1). The result suggests that Florida scrub jays may indeed avoid mimics after they have learned to avoid the sickening model.

Mimicry in which a relatively tasty prey mimics another poisonous species is called Batesian mimicry after its discoverer, the nineteenth century British explorer H.W. Bates. Another nineteenth century explorer (like Bates, in South America) was the German, Fritz Müller. Müller discovered another kind of mimicry, now called Müllerian mimicry, in which all the mimicking species are poisonous. All the species in a Müllerian mimic group benefit when an individual of any one species is eaten. The most extraordinary development of Müllerian mimicry is that of South American butterflies in the genus *Heliconius*. Two species, *H. erato* and *H. melpomene*, are Müllerian mimics. In any one place the two species resemble each other, but in different places they have formed "geographic races" (Figure 6.7). The butterflies look different in different places, but the two species always change in the same way. Thus two individuals of different species from the same place look more alike than two individuals of the same species from different places. In any one

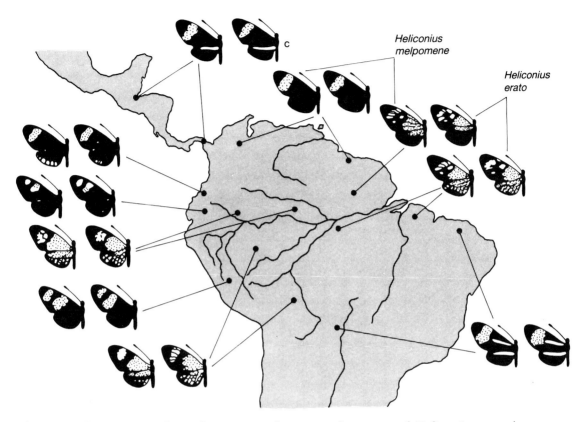

Figure 6.7   The two butterfly species *Heliconius melpomene* and *Heliconius erato* form remarkable parallel mimicry "rings" in Central and South America. In any one place the two species mimic each other, looking much the same, but in different places the members of a species differ: both species vary geographically in the same manner. *Heliconius* butterflies are poisonous to birds. *(After Turner)*

place, the two species gain by resembling each other, because predators will treat them both as the same kind of prey; few predators move far enough for there to be any advantage to the *Heliconius* in looking the same in distant places. All that matters is to educate the local predators, and that can be done with different colors in different places. The particular color pattern of a *Heliconius* presumably does not matter much, so long as it is memorable to the local birds. Their color pattern will then protect them from predation.

6.8   Fish schooling

In many species of fish, individuals swim together in large groups called schools. The habit of moving around in large groups is much more common in fish than in other kinds of animals. Within fish, it is more common among species that are small in size and active swimmers; even within a species, small individuals may school whereas large ones do not. Small tuna, for example, swim in schools, but when they grow large they become solitary. Very few of the large predatory species of fish swim in schools, though barracuda are an exception. The observation that large, predatory species of fish do not usually swim in schools might lead us to suggest that the advantage of schooling is in defense against predators, for large predatory fish are not themselves subject to predation.

Schooling might offer any of several kinds of protection. One is that a predator is less likely to detect fish prey if they form a school than if they spread out. There is an analogy here with ships in wartime, which sail in convoys to reduce the numbers lost to submarines. If all the ships sail separately it is more likely that at least one ship will stray into the field of view of a submarine than if the ships sail in a tight convoy; a convoy may slip by without being detected. Even once a predator has found a school, the prey fish are better off in the school than they would be alone, as has been proved by the following experiment. A prey species, such as the bleak or dace, is put in tanks either singly or in groups of up to 20; the groups will form schools. The experimenter then puts a predatory species of fish, such as the pike, into each tank. The result is that the predator catches more prey per unit time when feeding on single fish than when feeding on schools. The reason is that in a school the predator is confused by the multiplicity of moving prey items: it cannot concentrate on one without being distracted by another. The pike is perhaps in a similar difficulty as a human who is trying to hit a moving tennis ball when someone throws a second ball across his or her visual field: it makes it more difficult to hit the ball. The predator, moreover, has to concentrate when there are hundreds of whizzing prey. Another experiment supports the same theory. If predators find it difficult to concentrate on prey in schools, then they should become more efficient if some members of the school are made easier to concentrate upon. If an experimenter makes a few fish in the school distinctive, the predator should take more notice of them. And, indeed, when India ink was daubed on a few minnows and then put among a school, the painted minnows were disproportionately taken by predators. It was easier for the predator to keep its eye on them: in the normal school the predator is visually confused.

Schooling fish also take active steps to reduce the chance of predation (Figure 6.8). Schooling sticklebacks and catfish, for example, close ranks on the approach of a predator. Other species deploy more spectacular "set piece" avoidance tactics. When a school is attacked, the fish may "explode" away from the point of attack, in a dramatically simultaneous

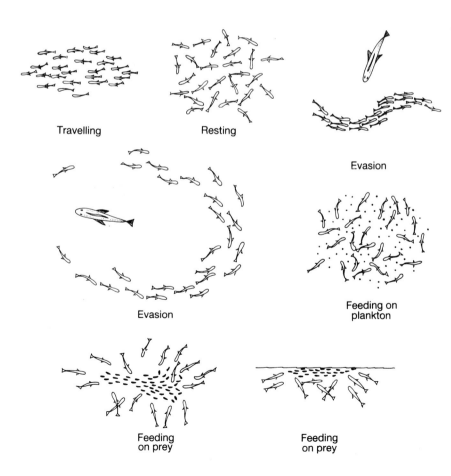

Travelling

Resting

Evasion

Evasion

Feeding on plankton

Feeding on prey

Feeding on prey

Figure 6.8

The arrangement of fish in a school depends on their environment. They align themselves differently according to whether they are traveling, resting, feeding, or avoiding being fed on. The school becomes less coordinated when the fish are resting or feeding. *(After Wilson)*

rapid movement. It may take less than a tenth of a second, yet the fish do not bump into each other. Another defensive maneuver is for the school to break in two in front of the predator, and then re-form after the two sides have swum around behind the predator.

Schooling, then, is a complex adaptation to avoid being eaten. It has other advantages as well. Fish swimming in schools move faster and use less energy than fish of the same species made to swim alone. The reason for this is uncertain. The Atlantic spadefish provide evidence for one theory. The spadefish swim faster in schools because of a slime given off by the swimming fish. The slime reduces the drag experienced by the fish swimming behind, which can consequently swim faster. Schooling fish thus swim faster as well as more safely than fish that swim alone.

We saw earlier, in the case of group-hunting carnivores, that living in groups can be an adaptation to increase the efficiency of hunting. It now appears that in other species the same habit can have an almost opposite function. Schooling in fish (and this is not the only example) serves to decrease the chance of being eaten. Group living has different functions in different species.

6.9   Summary

1. An animal's diet is the key to understanding its behavioral adaptations for feeding. Each of the five main large herbivorous mammals of the Serengeti has a specific diet, determined by its energetic needs and alimentary morphology and possesses feeding preferences and migratory responses to obtain it.

2. An animal must recognize and catch its food. The relatively inflexible toad recognizes food merely by the size of the dark moving objects in its visual field; its recognition mechanism is neurophysiologically understood.

3. Some species learn to recognize their prey after practice, which is called forming a search image.

4. Animal techniques of searching and catching food are fitted to their prey. Thrushes adjust their movements according to the spatial dispersion of their prey.

5. Lions, dogs, and wolves hunt down prey by cooperative group mechanisms. These ancestral mechanisms, in the case of dogs, have influenced the process of domestication.

6. Animals may avoid being eaten by active flight, as do moths escaping from bats; by camouflage, which requires behavioral adaptations to fit with the animal's appearance; by warning coloration, to teach predators to avoid sickening prey; by mimicry of successfully warning-colored species; or by aggregation in groups, such as the schools of fish.

6.10   Further reading

Colinvaux (1978) discusses feeding ecologically. Bell (1971) explains his work on the herbivores of the Serengeti, and Ewert (1987) describes his work on food recognition in toads. M. Dawkins (1971) gives the details of her experiment on search images; rat learning can be followed up through Garcia et al. (1989). For the experiment on thrushes the reference is Smith (1974a,b); search paths are part of the general subject of optimal foraging theory, which is reviewed by Stephens and Krebs (1986). See Schaller (1972) for lions; Vines (1981) was my source for sheepdogs. A further advantage of group living is information exchange about food, and Greene (1987) describes a remarkable instance in ospreys. Roeder (1965) describes his work on moths, Brower (1968) the unpalatability of monarchs to jays, Turner (1977) the subject of mimicry, and Edmunds (1974) discusses defensive adaptations as a whole.

CHAPTER 7

Signals

After looking at the question of how signals should be defined and recognized, we consider bird song and pheromones (chemical scent signals) as examples. We then move on to the famous dance language of honeybees, both as a remarkable signaling system and as a classic case study in the experimental decoding of a signal. We finish by discussing how signals evolve: we ask what the precursors were that signals historically evolved from and what the selection pressures are that shape the form of signals.

7.1 Principles of communication

The subject of the remaining four chapters is social behavior. Social behavior is mediated and organized by communication, and before we come on to such social, and antisocial, topics as fighting, sex, and cooperation, we should consider the means by which these interactions are controlled. We should discuss the principles of animal signals.

Students of communication never have been able to agree on the correct definitions of communication and signaling. A definition along the right lines but too broad, is that an animal has signaled when it changes the behavior of another animal. The definition can be made more accurate, but not wholly satisfactory, by specifying that the other animal must have changed its behavior because it perceived the signal through its sense organs, and was not physically forced. On the former definition, pushing someone in a river would have to be called a signal because it certainly would change their behavior, but this would be excluded by the definition requiring the influence of the signal to be mediated by the recipient's sense organs.

The criterion of behavioral change is necessary in order to recognize signals by external observation. Most signals have been discovered by simply watching the behavior of interacting animals. If members of a species consistently perform an activity, such as running away, after other members have performed another recognizable activity, such as baring their teeth, then teeth-baring is probably a threat signal. The evidence of simple observation is not, however, perfectly convincing. It can only establish a correlation between two behavior patterns, but a correlation can always be explained by both activities being caused by a third, unobserved activity. Experimental evidence is therefore more convincing. Tinbergen, for example, presented model bills of adult herring gulls (*Larus argentatus*) to the chicks of that species; the chicks responded by pecking at the red spot at the top of the bill, which strongly suggests that the red spot is a signal meaning "peck here" (Figure 7.1). (The pecking of the spot is the chick's signal method of asking for food: if the spot were on its parent's, rather than a model bill, the chick would receive a meal.) It is difficult to believe that every time Tinbergen presented a model bill to a chick his activity coincided with a third, unobserved variable that was really signaling to the chick to start pecking; that would, however, be possible for the natural observation that chicks peck when their parent arrives. Scientists therefore require experimental evidence to test between causes and correlations. Now that we provisionally have fixed what a signal is, and how one may be recognized, let us consider some examples of signals—the songs of birds, the pheromones of moths and ants, and the dance of honeybees—before we consider the theoretical question of why signals have evolved in the form that we see in nature.

7.2   Bird song

Birds, like mammals, produce sound by blowing air from their lungs over vibratory vocal cords in the trachea, although in birds the vocal cords are situated slightly closer to the lungs than in mammals. Bird song is a very familiar kind of animal behavior. It has been celebrated by poets and enjoyed by most people. It is also particularly easy to study, because sound can be recorded and reproduced by a tape recorder. So, why do birds sing? And why do they sing in the way they do?

Male birds do most of the singing, which gives an immediate clue that singing has something to do with sex. In fact, singing seems to serve two main functions in birds: defending territory, and attracting and stimulating females to mate. The following experiment shows the importance of song in territorial defense. John Krebs removed the resident pairs of great tits (*Parus major*) from their territories in Wytham Woods near Oxford (U.K.). Some of the territories he left empty; in others he placed loudspeakers broadcasting the song of a great tit. He then watched to see how long it took another pair to occupy the two kinds of territories. It took longer for the territories with loudspeakers to be occupied than it

Figure 7.1   Experiments are needed to confirm that any particular structure or behavior pattern functions as a signal. Herring gull (a,b) and black-headed gull (c) chicks naturally peck at the tip of their parents' bills to beg for food. At the tip of the gull's bill is a red spot that was believed to be the signal to the chick to peck. Tinbergen presented various (d–f) more or less realistic model gull bills, with different colored spots (compare e and f) to gull chicks. He measured the rate at which they pecked the different models, and confirmed that the red spot on the tip of the bill acts as a signal to the chick to peck, in order to beg for food. *(Photos: Niko Tinbergen)*

Figure 7.2

An experiment shows that song in great tits functions in territory defense. (a) Territories' boundaries before the experiment. (b) All territorial male great tits were removed from the area, which was then divided into three regions: in the experimental area loudspeakers broadcast the songs of male great tits and two control areas were either silent or had loudspeakers broadcasting the tune of a great tit played on a tin whistle. (c) The course of reoccupation by intruding great tits: the experimental area was occupied more slowly, presumably because the song deterred intruders. *(After Krebs)*

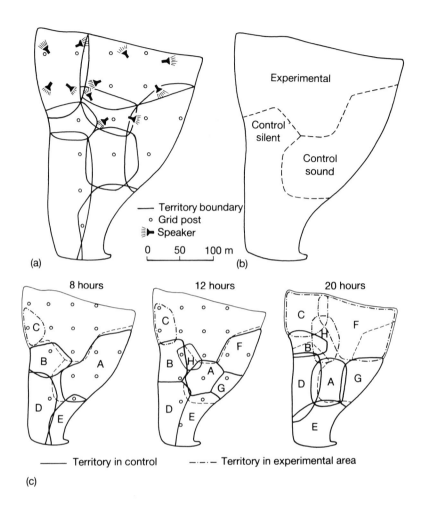

did for the silent territories (Figure 7.2). He then took the experiment a stage further. Great tits sing a repertory of one to eight distinct songs. Why do males sing so many songs? What can eight songs do that one cannot? As before, Krebs removed pairs from their territories and put loudspeakers in instead. This time the loudspeakers in some territories broadcast one song, whereas the loudspeaker in other territories broadcast a repertory of up to eight songs. It turned out that the territories in which a larger repertory was broadcast took longer to be reoccupied. For instance, it took 14 hours more for a territory with eight songs to be reoccupied than one with only one song. In the great tit, then, males sing to keep intruders out of their territories, and a repertory of songs does so more effectively than a single song.

The songs of male birds may also stimulate female reproduction. Don Kroodsma experimentally played normal and artificially reduced repertories of male canary song to female canaries. The females who were played the reduced repertories turned out to build nests at a lower rate. Another, less direct observation that suggests that song serves to stimulate females is that males in polygamous species of North American wrens have more complex songs than have closely related monogamous species. Competition among males for mates is stronger in polygamous than in monogamous species. If songs have been evolved by males because of competition for mates, we should expect males of polygamous species to have more complex songs than males of monogamous species.

"Alarm calls" are another kind of sound made by all birds. Individuals of many bird species give alarm calls when they spot a dangerous predator. A blackbird (*Turdus merula*), for example, might give an alarm call on seeing a hawk flying overhead. The alarm call stimulates other nearby blackbirds to take evasive action. As was first noticed by Peter Marler, the alarm calls of many species sound similar (Figure 7.3). They all share certain acoustic properties which, he thought, would make the call difficult to locate. Making a noise when a predator is nearby is, after all, dangerous; it may attract the predator to the noisy bird. There would be an advantage in giving an alarm call that is difficult to locate. An experiment of Charles Brown has confirmed Marler's hypothesis. Brown measured the accuracy with which horned owls (*Bubo virginianus*) and red-tailed hawks (*Buteo jamaicensis*) oriented to recordings of mobbing calls (which are not acoustically camouflaged) and alarm calls of the American robin (*Turdus migratorius*). He found the predatory birds oriented less accurately to the alarm calls, as Marler would have predicted.

7.3 Pheromones

Pheromones are chemical signals released by one individual and smelled by another, whose behavior is influenced. They are smell signals. Hu-

Figure 7.3
Many species of birds give alarm calls when they see a dangerous predator. The particular bird giving the alarm call runs the risk of attracting the predator to itself, so the alarm calls have acoustic properties that make them difficult to locate. On the left are the alarm calls of four species of songbirds; their similarity is suggested by comparison with the ordinary song of the four species, shown on the right.
(*After Marler*)

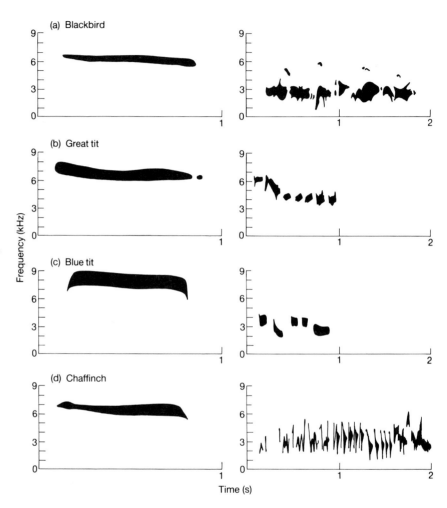

mans, as it happens, make little use of smells in communication, but in some other species, such as ants, it is highly important.

The best known pheromone is emitted not by an ant but by the female silk moth (*Bombyx mori*). The pheromone is called bombykol, and it attracts male silk moths. We humans cannot smell bombykol, but it is potent in its effect on male silk moths. They can smell the bombykol released by a single female at a distance of over a mile. The male's olfactory organs are his antennae, which are enlarged and have a fine mesh of side branches; they are so sensitive that, in experiments by Dietrich Schneider, a single molecule of bombykol elicited a response in the sensory neuron of the male. Only about 200 molecules need to hit the antennae of the male for its brain to issue commands for flight. The antennae also show a specific response: they sense only bombykol, and are not stimulated by very similar molecules. The response of the male silk moth to bombykol is an imperfect system to understand pheromones because the male silk moth is flightless, probably because of long domestication. It retains its receptivity to bombykol probably because it was useful in its ancestors, which presumably resembled most moths in that the male flew toward the female when they sensed the pheromone.

Once a male moth has sensed female pheromone, his task is to fly in the correct direction: he must make the correct taxic response (p. 64). He could achieve this by either of two techniques. He could fly around, measuring the pheromonal concentrations, and then orient in the direction where the concentration increased; or he could simply turn upwind. In fact, the evidence suggests that male moths follow the second rule: on sensing pheromone they simply fly upwind. If at any point they lose the scent they fly in zigzags from side to side until they catch it again, and then fly off upwind again. That set of responses is enough to guide them to the female.

Female ants also release pheromones to attract males; their pheromones, however, are only effective over distances of 25–30 yards. The social behavior of ants is controlled mainly by pheromones: they do use visual and auditory signals, but most of their signaling is by means of smells. Not only mating is controlled pheromonally—so too are finding and exploiting food, recruiting nest mates for battle, and warning about enemies. The collection of scent glands employed by an ant such as *Iridomyrmex humilis* (Figure 7.4) beats the range of mere human cosmetic collections, and all the ant's scents are meaningful.

Let us consider how several ant species recruit nest mates for group action. If, for instance, a lone ant finds a food source too large for it to bring back to the nest by itself, it runs back to the nest, leaving a pheromone trail on the way. Different kinds of ants leave trails from different pheromone-releasing organs. *Solenopsis,* for instance, releases

Figure 7.4

Pheromones are manufactured and released from special glands. Here are the 12 main pheromone-releasing organs of the ant species *Iridomyrmex humilis*: (1) mandibular glands, (2) maxillary gland, (3) thorax labial gland, (4) poison gland, (5) vesicle of poison gland, (6) Dufour's gland, (7) postpharyngeal gland, (8) metapleural gland, (9) hind gut, (10) anal gland, (11) reservoir of anal gland, (12) Pavan's gland. (*After Wilson*)

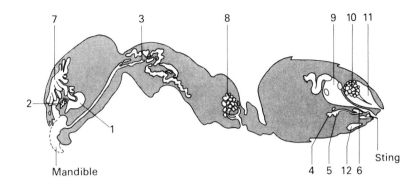

its trail pheromone from its Pavan's gland, *Myrmica* from its poison gland, and *Lasius* from its rectal gland. When the food-finder arrives at its nest it uses further pheromones to recruit other ants to come and collect the food; *Myrmica rubra*, for instance, having laid its trail from its poison gland, attracts its nestmates back to the food with a pheromone from its Dufour's gland. Recruitment is not always effected by mass-acting pheromones. The food-finder might instead single out one other nestmate, and then lead it alone back to the food source. They match the recruiting party to the size of the food source. Bert Hölldobler has discussed how *Leptothorax* workers may recruit and lead a single ant when the food source is too big for one ant but only needs two to move it. When, for instance in *Solenopsis*, the food source is so large that many ants are needed, mass-acting pheromones are released.

The scent trail of the ant is released as a liquid (which evaporates) on the ground, but it is known that it is perceived by scent rather than taste. In an experiment, leaf-cutter ants of the species *Atta texana* had to follow a trail by walking along a plastic roadway placed just above the trail; they followed the trail as usual, but must have done so by the airborne odor of evaporated scent. Like the silk moth, ants sense pheromones through their antennae, but they make continual use of both antennae to keep them in the right direction. Unlike the moths, which, we have seen, ignore relative pheromonal concentrations and simply fly upwind, ants steer by balancing the pheromonal concentration to the right and left. If the concentration increases to the right they turn in that direction until both antennae are sensing equal concentrations: this rule will guide them straight down the odor trail. In an experiment, Hangartner placed two trails in parallel; to begin with they were of equal concentration, but as they proceeded one of the trails became progressively weaker. The ant species *Lasius fuliginosus* started by walking between the two trails, and then moved across to the stronger trail in such a way as to balance the odor strengths sense by their two antennae (Figure 7.5).

Recruitment pheromones are not confined to the exploitation of food: they are also used to recruit armies for territorial disputes, or (in certain species) for slave raids. Not all ant species enslave other ants, but *Formica subintegra* (for example) does. It raids the pupae and larvae from the nests of other ant species, brings them back to its nest, and when the pupae and larvae hatch they work as slaves for their captors. These slave raids are coordinated by a pheromone from the ant's Dufour's gland (which is very enlarged in this species—Figure 7.6). E.O. Wilson calls this pheromone a "propaganda substance."

With this kind of warfare going on between nests, the ants need to be able to distinguish their own nestmates from members of other nests. This distinction is made possible by other pheromones, called colony odors. Each nest has its own distinctive smell (see also section 10.4, p. 249).

Ants warn their nestmates about enemies with alarm pheromones. On scenting an alarm pheromone, an ant may do any of a variety of things: it may run away from the source of the scent, it may freeze and "play dead," or it may run toward the source of the scent and attack any nearby enemies. The different responses are probably stimulated by different alarm pheromones. The diversity of messages in an alarm pheromone is suggested by a chemical analysis carried out on the weaver ant *Oecophylla longinoda*. Its alarm pheromone contains over thirty different chemicals. The effects of only four of these have been tested. One makes the recipient ant generally "alarmed," another makes it run

Figure 7.5
Ants follow pheromone trails by balancing the concentrations smelled through their two antennae. These *Lasius fuliginosus* are walking along two experimental trails: the bottom trail has constant concentration; the concentration of the top trail starts the same as the bottom one but gradually decreases. The ants initially walk between the two but cross over to the bottom one at the point predicted if the ants are balancing the concentrations measured on either side.

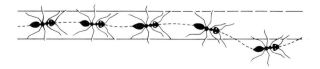

Figure 7.6
The slave-making ant species *Formica subintegra* (top) has an enormously enlarged Dufour's gland compared with a more typical ant species such as *Formica subserica* (bottom). Dufour's gland is the source of the pheromones, called "propaganda substances," used to confuse the defending workers in the nest that the slave-makers raid. *(After Wilson)*

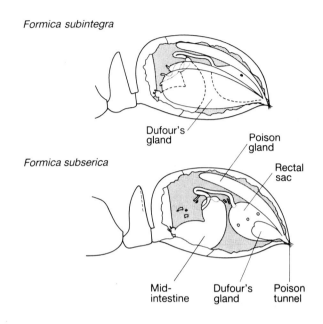

toward the source, and two others make it bite. The alarm pheromone of the weaver ant is a complex message that elicits a whole series of responses in its nestmates.

7.4   The dance of the honeybees

7.4.1 The meaning of the dance

If you put out a dish of syrup during the summer, it might be days until the first bee finds it. Bees treat a new dish of syrup like a new flower, recently opened and full of nectar: its fortunate discoverer feeds on the syrup, and then flies home to her hive. There will now be a much shorter time delay until the next bee comes and during the next few hours many bees, all from the same hive as the original discoverer, will visit the food source. The reason for this sudden rush of bees is that the first bee tells her hivemates about the syrup. She tells them where it is, how far, and in what direction; she tells them what the food tastes and smells like; and she also tells them how good a food source it is. All of this is disclosed in a special "dance," the discovery and translation of which is one of the great achievements of modern ethology. It was worked out by the Austrian biologist Karl von Frisch in the middle of the century, by methods we shall discuss in the next section. But first, how does the dance work? Inside the hive, where the dance is performed, it is dark and the bees are moving around on the vertical wall of the honeycomb. The bee that has found a food source does one of two main kinds of dance according to how far away the food source is. If it is near the nest she does a "round" dance, in which she walks around in circles, reversing direction every turn or two (Figure 7.7a). The round dance does not tell the recruits the direction to food, but it does tell them that food is nearby. If the food is further away, the bee does a "waggle" dance (Figure 7.7b). The distance from the food source to the hive at which the changeover from round to waggle dance takes place differs between different races of bees; *Apis mellifera lamarcki* individuals, for example, do round dances up to distances of only about 4 yards, whereas bees of the race *Apis mellifera carnica* do it up to about 16 yards. There is no possibility of misunderstanding between dancer and spectator, however, because the changeover distance is constant within any one hive. The waggle dance is a "figure eight," and it tells the direction as well as the distance to the food. The key part of the dance is the run up the center, between the two circles of the eight. While running up the center the dancer waggles her abdomen from side to side. It is the angle of this part of the dance, with respect to the vertical, that tells the other bees the direction to the food. They understand straight up as meaning the direction of the sun, and they know which direction is straight up because they are sensitive to gravity. The angle of the waggled part of the dance with respect to vertical symbolizes the angle between the food source and the sun. Therefore if the bee orients her waggle 90° to the left of vertical, the food source is 90° to the left of the sun. The bee tells her audience how far it is to the food by another property of the dance, its rate. If she takes longer to complete a circuit, she means the food is further away. Again, the precise relation between the rate of dancing and the distance to the food differs between races. For example, a rate of seven circuits per second means a distance from hive to food of about 400 yards to an *Apis mellifera carnica,* 330 yards to an *Apis mellifera mellifera,* or 260 yards

Figure 7.7
The dance of the honeybee tells the location of food sources in relation to the hive. (a) The "round" dance, which says that food is nearby, but does not show the direction to it. It is used when the food source is so close that the recruits will have no difficulty in finding it without being told the direction. The dance consists of a series of clockwise and counterclockwise circles; the top bee in the picture is performing the dance. (b) The "waggle" dance, which says both the distance to and direction of a food source. The distance is told by the rate of the dance and the direction by the angle of the center part of the dance, during which the dancer wags her abdomen: the angle of this part of the dance relative to the vertical is the angle of the food source relative to the sun.

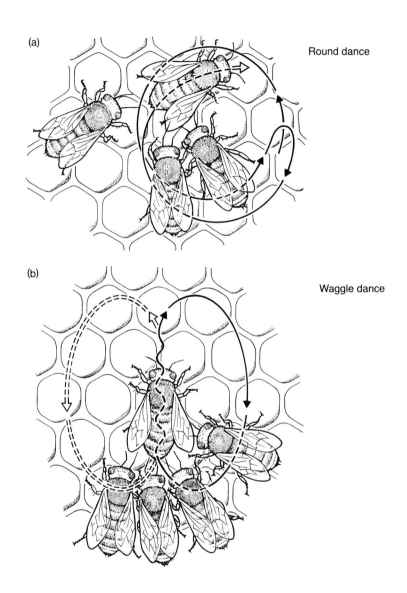

(a)   Round dance

(b)   Waggle dance

to an *Apis mellifera ligustica*. Scent also matters. The scout bee sometimes marks the food source with her scent. During the dance the other bees can smell her scent and then use this knowledge in finding the food. The dancing bee also regurgitates food to the recruits to illustrate its taste.

Bees can use the dance to point to things other than food. They also dance about possible new nest sites, and about water sources when water is needed to cool down the nest. The dance is a means of indicating distances and directions in the environment outside; it thus enables the nest, in several respects, to exploit that environment more efficiently.

7.4.2 Decoding the dance language

Karl von Frisch (1886–1982) spent his life discovering unsuspected sensory and behavioral skills in animals. We have seen how von Frisch proved von Hess wrong about the sensitivities of fish (p. 68). Soon after that feud, von Frisch made his even more controversial discovery of the dance language of the honeybee. We have just discussed its form: let us now consider the evidence that led von Frisch to the discovery.

In the 1920s, when von Frisch was carrying out similar experiments on honeybees to those he had done with fish, he noticed that although initially a lone bee came to his food dish, soon afterwards many bees came. He noticed too that the lone bee that first found the food performed a regular sequence of movements on returning to her hive. At the time he thought this "dance" merely alerted the other bees to the presence of food, which they then located by smell. The bees could learn the smell of the food from the sample regurgitated by the discoverer. Von Frisch did not doubt the odor theory of how honeybees find food until the 1940s. Two kinds of experiments then led him to think again; in their most advanced forms, they are called the "fan" experiment and the "step" experiment. In a step experiment (Figure 7.8) von Frisch trained a bee to come to a dish of scented, sugary water. While the bee was back in its hive dancing about the source, von Frisch put out other dishes containing the same scented sugary water nearer to and further from the hive, but all on a line with the original dish. If the bees found the food by smell they would go to the nearest dish, but in fact they went to the original dish which the first bee had been to. The idea is similar in a fan experiment (Figure 7.9). In the step experiment, other dishes are put in line with the original dish. In a fan experiment, other dishes are put at the same distance as the original dish, but in different directions. Again, all the dishes contained the same scented food, but the bees mainly went to the dish where the first bee had been. Meanwhile, von Frisch had looked more closely at the honeybee dance. He noticed that he could predict the

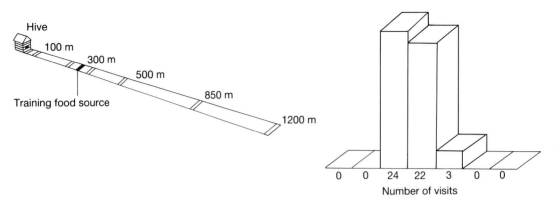

Figure 7.8 The "step" experiment of Karl von Frisch suggested that bees can tell the distance from hive to the food source that was the subject of a waggle dance. He trained a bee to come to a particular source (black in the figure) and then put out seven similar food sources at various distances from the hive. The bees mainly came to the source at, or just less than, the distance of the original source.

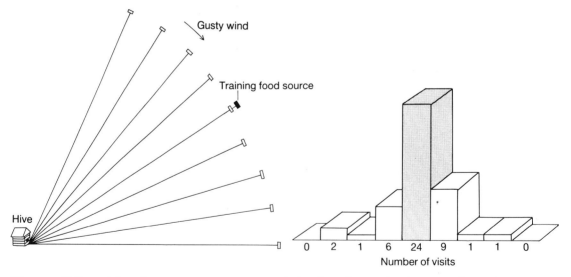

Figure 7.9 The "fan" experiment suggested that bees know the direction from their hive to the food source that has been danced about. Von Frisch trained a scout bee to come to a source and then placed out new sources at equal distances from the hive, but only one at the same direction as the original. The recruits mainly came to the site of the original food source.

direction to the food source from the dance. As he varied the position of the food source, the dance varied in a predictable way. He concluded that the bees signaled the direction to the food by the dance.

Von Frisch's theory initially was not believed. However, through the 1950s and early 1960s it became widely accepted by biologists. It was not challenged seriously until the late 1960s, when Adrian Wenner and others pointed out that von Frisch had not excluded the possibility that the returning, dancing bee brings back with it a smell that is specific to the locality where the food is. A series of inconclusive experiments failed to settle the issue until the 1970s, when J.L. Gould finally confirmed von Frisch's theory. Gould made use of the fact that in the dark the bees orient their dance to gravity but if they can see the sun they dance directly with respect to it. They also will orient the dance to an electric light bulb, as if it were the sun. In his experiment, Gould painted over the ocelli (small light-sensitive organs) of some of the members of a hive, which would then dance as if it were dark but otherwise behave normally. He trained the bees whose ocelli had been painted to a food dish, and then let them dance to their unpainted hivemates under an electric light bulb. The dancing bees danced with respect to gravity; the spectators, whose ocelli were unpainted, interpreted the dance as being with respect to the light bulb. If von Frisch's critics were right, and the dance was irrelevant, the bees should find the food as usual. If, however, the dance told the bees where to go, they should go off to the wrong place (depending on where Gould had put the light bulb). They went to the wrong place. Gould, moreover, could exactly control where the bees would go to by moving the light bulb. His experiments show that the bees must be taking notice of the dance, and confirmed von Frisch's theory more strongly than before.

7.5 The evolution of signals

7.5.1 The evolutionary origins of signals

Why do animal signals have the properties that they do? One influence is the environment in which the signal is naturally broadcast. Different kinds of signals work better in different environments. Signals that use reflected light, for instance, cannot be used in the dark. Likewise, light does not penetrate through water as well as it does through air, whereas with sound and chemicals it is the other way around. Animals communicating in the dark or through water are therefore more likely to use sound or chemical signals; terrestrial, diurnal animals make more use of light signals. A second influence is ancestry. Signals usually resemble other activities, performed for other reasons, in the behavioral repertory of an animal. It is likely that the nonsignaling activity existed before the signal, and the signal was derived evolutionarily from it by gradual modification.

Three particular classes of activities have been identified as the most common sources of signals: intention movements, displacement activities, and activities controlled by the autonomic nervous system. Intention movements are activities that tend to precede some other activity, for more or less inevitable physical reasons. A familiar human example is the habit of sitting up, or looking toward the door, when you wish to leave the room. You have to get out of the chair before you can leave, and that requires certain physical movements; looking toward the door is less inevitable, but it is still sensible to check that there are no obstacles in the way. Given the usual circumstances of these actions, they are good precursors of signals. If one person wishes to signal to another that it is time to leave, he or she may more elaborately look at the door, or sit up. Then the intention movement is being exaggerated into a signal.

An analogous process probably took place in the evolution of many animal signals. The "upright" signal in the gull, for example, is probably derived from an intention movement. Upright is a threat signal, in which the gull stands upright and points its bill downward and toward the other gull it is threatening (Figure 7.10). This is very similar to the posture a gull physically must take up just before attacking another gull by jabbing it with its bill or biting it. A gull cannot jab downward unless it first assumes the upright posture. When ancestral gulls saw another gull nearby in the upright posture, it was advantageous to run away. Natural selection favored gulls that ran away rather than staying and being pecked. The upright posture was then starting to evolve into a

Figure 7.10
The aggressive "upright" display of gulls is illustrated here by a black-headed gull. *(After Tinbergen)*

threat signal. Once a number of individuals in the population tended to flee from upright gulls, selection favored aggressive gulls that assumed the upright posture in a particularly clear manner that was unlikely to be mistaken. It had then evolved into a pure threat signal rather than just being the first stage of an attack.

The reason why displacement activities and autonomic nervous responses should evolve into signals is less clear. Indeed, the function of displacement activities themselves is uncertain. They are actions that appear irrelevant to the circumstances, and are generally performed at times of motivational conflict. A bird, for example, on seeing a conspecific nearby may be simultaneously inclined both to approach it and to flee from it: a motivational conflict between attack (or courtship) and flight. It may then, of all things, preen its feathers, an activity relevant to none of its options. Now, in the courtship of many species of ducks (see Figure 2.4, p. 26), the male points his bill toward his back feathers in what looks like a stylized modification of preening. For this reason, Tinbergen suggested the courtship signal was ancestrally a displacement activity. (Displacement activities can be expected to appear at the time of courtship, because it is a time of great motivational ambivalence.)

The facts are not in doubt, but they are difficult to make sense of because the function of displacement activities is obscure. If displacement activities really are irrelevant, why do animals perform them? Why should an ambivalent bird preen its feathers (and why do we scratch our heads when confused)? Perhaps these actions are, in some manner still unknown to us, not irrelevant; the judgment of irrelevance is that of the human observer, not the animals concerned. One possible reason why signals have evolved from displacement activities in courtship is as follows. In nature, female ducks have to distinguish purely aggressive male ducks from those that are more willing to mate. The former kind of male duck will not be ambivalent and will not perform displacement activities when near a female duck; the latter kind are more likely to be ambivalent (and so to perform displacement activities such as feather preening). Thus female ducks will be selected to prefer to mate with male ducks that clearly signal they are unaggressive, and displacement activities will evolve into courtship signals. (We return to the topic of courtship in Chapter 8.)

The third source of signals, autonomic nervous activities, was first noticed by Darwin, although their nervous control was not then known; his example was shivering at times of fear in humans. The autonomic nervous system was not discovered until after Darwin's work. It is a system of nerves, separate from but connected with the central nervous system, and it controls most of the unconscious activities of our body, together with the responses of arousal, rapid breathing, sweating, and so on, that are characteristic of the "fight or flight" reaction. Desmond Morris later drew attention to several signals that resemble activities produced by the autonomic nervous system, such as auditory signals (shouts) that may have evolved from rapid automatic breathing, and even the scent-marking by urine at territory boundaries in mammals. The autonomic nervous system tends to cause mammals to urinate at times of fear. Urination could then have evolved into a territorial signal because an individual mammal will not be fearful when inside its territory, but as it moves away from its home site it will become increasingly fearful. At some point its autonomic nervous system will increase its probability of urination. That point in space is a good indicator of where that animal can advantageously be attacked: it is advantageous to attack him on your side of the border (where, as his autonomic response reveals, he is likely to run away), but not on his (where he is more confident). This is only one possible argument, and it is uncertain, but it shows how an argument about the evolution of signals can be constructed.

7.5.2 Natural selection and "ritualization"

In the last section we mainly considered the evolutionary precursors of signals. Between the signal's origin in ancestral animals, and its modern observable form, that signal underwent evolutionary change to make it perhaps more elaborate, or more distinct as a signal: it was evolving into a signal. The evolutionary process of transformation from a primitive precursor to a fully evolved signal is called "ritualization." Why does natural selection favor ritualization? One general reason is to make the signal stand out against background noise—to improve the signal-to-noise ratio, to make the signal more likely to be detected by the receiver. In the case of bird song, or any other acoustic signal, there is literally noise in the background: the noise of wind in the trees, of other animals, and so on. How can a bird make its song stand out against this background? One method is to increase the song's volume; there are other possibilities, such as using acoustic frequencies that are relatively unrepresented in the background noise, or by giving the song a melodic structure that differs from the background. Bird song has all these properties. The evolution of a song with acoustic properties that improve its signal-to-noise ratio is an example of ritualization.

In a case like the upright threat signal of the gull, we can imagine another kind of noise. The receiver gull here is assessing the likely future behavior of other gulls in its colony, and if it infers one is likely to peck it, it keeps clear. However, a gull colony is a confusing place. There are a large number of birds, moving rapidly in many directions and with many behavioral purposes. Most gulls most of the time are not attacking other gulls, and gulls will not allocate much of their time to assessing whether they are about to be attacked. At the early evolutionary stage of the upright signal, the upright posture was assumed for only a fleeting moment before the attack came. Most times, the victim would not have noticed it before it was pecked, and a fight would break out. There would not have been enough time to see it, with all the distractions of a gull's life. If the victim had seen it, it would have retreated and the attacker could have saved itself the costs of the attack—costs such as the risk of injury from retaliation (or, maybe, subtle long-term social costs). It is therefore advantageous to make the signal clear against the background of competing claims on the other gull's attention. Selection would favor a slowing down of the upright posture, in the manner described above, making it less likely to be overlooked. Again, the signal is being "ritualized," and the reason is to improve the signal-to-noise ratio.

7.5.3 The evolution of "honesty" in signals

Ritualization probably proceeds differently in signals for which the signaler and receiver have a community of interest and in signals for which their interests conflict. If there is a community of interest, signals are likely to be "honest": the signaler will only signal correct information about the external environment (such as where food is), its qualities (such as strength), or its intentions (such as its probability of attacking). (The word "honest" here refers only to whether the signal itself accurately reveals the state of the real world; it does not imply, as it might in a discussion about humans, that the signaling individual consciously intended to deceive the recipient.) The signal is also likely to be cheap, because the signaler will not have to persuade the receiver about the subject of the signal. For example, the dance of the honeybee, and many pheromonal signals in ants (though not "propaganda substances"), are probably signals in which there is little or no conflict of interest between signaler and receiver. Both are members of the same colony and (as we shall see in Chapter 10) natural selection favors cooperation within a colony. We should not expect bees to signal false information about food sources, sending hivemates off on spurious journeys about the countryside.

In other cases, such as threat signals, the interests of signaler and receiver may conflict. Signals are then more likely to escalate during evolution into highly elaborate and costly forms, and we can be less certain that signals will be honest. Consider a gull signaling something equivalent to "I am stronger than you" to an opponent in a fight. If the other gull simply backs down on receiving the signal, natural selection clearly favors the individual who gives the signal, whether or not he or she really is stronger. A "dishonest" signal could be favored. However, in a population of gulls who all signaled "I am stronger than you," natural selection would favor individuals who did not back down, and who used other indicators of strength in an opponent. Selection would then favor individuals who faked those indicators too. We might therefore expect that evolution would proceed in an unending "arms race," as signalers evolved to deceive receivers and make them behave in the interests of the signaler. Some signals may indeed have evolved in that manner.

But the arms race may not continue forever. If a signaler gives an honest signal of its strength, selection favors receivers who treat it as such. Evolution will reach an equilibrium, provided that weaker individuals are not selected to fake the signal of strength. This might be for either of two reasons. One is that the signal might simply be impossible to fake, perhaps because the signal is physically directly related to fighting strength. Depth of sound is proportional to body size, for example, and it may be almost impossible for a small animal to produce a deep sound. Then sound depth could evolve as an honest signal of strength.

The second reason is more general. Honesty may be favored if the signal is costly to produce, though a special condition is required. The signal of strength must cost less for a really strong individual to produce than for a weak individual. Then a strong individual can gain more from giving the signal of strength: although the weak individual could fake it, the benefit of having opponents back down would not outweigh the cost of producing the signal. In the case of a signal about fighting ability, the signal might actually reduce the ability to fight. A stronger individual could then afford to give a more energetic signal than could a weak individual, and it would still be more likely to win any fight that did take place.

We could make an analogy with signals of wealth in humans. Wealth can be displayed by potlatchlike acts of conspicuous consumption. The signal itself reduces the signaler's wealth. However, a rich person can put on any particular level of display at less cost to himself or herself than can a poor person. Thus a signal of wealth that itself conumes wealth is likely to be a reliable indicator of wealth. A cheap verbal statement (of the form "I am richer than you") is unlikely to be reliable.

These are mainly recent theoretical ideas and have yet to be convincingly tested. However, they do make sense of a number of features of signals. Signals where there is a conflict of interest do often seem to be more elaborate and energetic (i.e., more costly to give) than signals where there is a community of interest. Moreover, signals do often reduce the quality being signaled: we shall meet, at the beginning of the next chapter, a signal of strength in fights between deer that does reduce fighting ability. Also, the theoretical arguments suggest how ritualization may proceed very differently in different signals, and what properties a signal must have if it is to be "honest" where the signaler and receiver have a conflict of interest.

7.6   Summary

1. A signal by one animal indirectly leads another animal to change its behavior.
2. Signals are mainly detected by observing behavior to see which activities predictably lead to changes in the behavior of others, but this kind of evidence is philosophically unconvincing, and is best supplemented by experiment.
3. The form of signals can be understood in terms of the medium in which they are transmitted, their ancestry, and the process called ritualization by which they evolve from their ancestral form into signals.
4. Male great tits sing in order to warn other males away, as can be shown by leaving a loudspeaker playing great tit song in an unoccupied territory; more complex songs are more effective warnings. Male canaries sing at least in part to stimulate females to prepare for reproduction. Many birds give alarm calls to warn of danger, but it is not certain whether the acoustic properties of the alarm calls make them difficult to locate.
5. Chemical signals are called pheromones. Male moths fly toward females when the female releases a pheromone; ants use pheromones for many purposes—recruitment, propaganda, and warning of danger.

6. Honeybees tell hivemates the direction, distance to, and nature of, new food sources by special dances. Von Frisch initially decoded the dance language by his "fan" and "step" experiments. The experiments, however, did not control against "local" odors. Further experiments were needed to test whether bees used the dance or local odors to find the food source. These experiments confirmed that the dance language is used.

7. Ritualization may proceed differently according to whether the signaler and recipient have conflicting interests. If they are cooperatively exploiting a resource (as are honeybees), signals will convey accurate information. If they are competing, signalers may be "dishonest," or they must have particular properties that prevent dishonesty.

7.7   Further reading

Cullen (1972), Marler (1959), and Wiley (1993), among others, discuss the principles of communication. A special issue of *Philosophical Transactions of the Royal Society of London* (1993, series B, vol. 340, pp. 161–255) contains several papers on the topic. Morton and Page (1992) discusses acoustic signals, including bird song. Agosta (1992) describes pheromonal signals; Schneider (1974) explains how male silk moth pheromones are received, and Wilson (1971) gives an account of the use of pheromones by ants. Stoddart (1990) discusses scent in humans. Von Frisch (1967) is the classic account of the dance language of the honeybee, but it does not cover the recent controversy: see Gould (1976), and Wenner and Wells (1990) for a different viewpoint. Michelsen et al. (1989) describes an experiment in which a mechanical model of a dancing bee successfully recruited bees. See Grafen (1991) and Maynard Smith (1991) on signal honesty. Cheney and Seyfarth (1990) and Seyfarth and Cheney (1992) describe their superb study of signal meaning in vervet monkeys, a topic also discussed by various authors in Whiten (1991).

CHAPTER 8

Sexual Behavior

Animals spend an apparently inordinate amount of time and energy in courtship, and we begin by asking why. One incomplete answer is the need to find a mate of the correct species, but we move on to Darwin's theory of sexual selection for a richer set of explanatory possibilities. We look at how sexual selection operates, and some evidence for it. We finish by considering why different species have various "mating systems"—why some are monogamous and others polygamous, and why parental care is sometimes performed by males, and sometimes by females.

8.1   Introduction

Sexual behavior poses a set of related questions about mating and the rearing of offspring. Before conception, there may be competition among males, and a characteristic (often strange) sequence of behavior patterns in which the sexes court each other. The term courtship, as used by students of animal behavior, refers to all the behavioral interactions of the male and female that come before, and lead up to, the fertilization of eggs by sperm. Its form and ostentation vary among species. In some species it does not even exist. In others, such as the stickleback (which we shall consider later), it lasts a few minutes. In others it may last for months. Male and female waved albatrosses (*Diomede irrorata*), which live on Isla Española of the Galapagos Islands, may court each other, with an extensive repertory of stereotyped movements of the neck and bill, for several hours a day, day in and day out for much of the year (Figure 8.1). The question courtship poses is why it exists, and why, in some species, it has taken on so extravagant a form.

One answer is that courtship ensures that animals mate with other individuals of the correct species, sex, and condition. Its extreme development, however, of numerous energetic and colorful displays, and extreme structural modification in males, cannot be so easily explained. It is difficult to believe that all the paraphernalia of sex is needed merely to ensure that males are not confused with females or males of another species: that could be achieved much more quickly and less colorfully. Moreover, the most bizarre sexual traits, such as the amazing plumage of male birds of paradise, probably decrease the chances of survival of their bearers. They use up energy, hamper flight, and attract predators. They are therefore something of a puzzle. They must possess some hidden function to compensate their obvious disadvantages, because if they did not they would have been eliminated by natural selection. Darwin invented a special theory to solve the problem, his theory of sexual selection. We shall come back to that.

After fertilization one or both of the parents may look after and help to rear the offspring. The question of why parents should look after their young is part of the general question of altruism, which is the subject of the next chapter. But the sexual division of labor—whether it is the male, the female, or both parents who look after the young—is controlled by much the same forces as are other sexual differences and is therefore appropriately treated in this chapter. We shall start by confirming that sexual behavior does indeed ensure that mates are of the correct species, and then see how well Darwin's theory, modified and tested by recent work, can explain the full variety of sexual behavior in animals both during courtship and later, in parental care.

Figure 8.1   The courtship of the waved albatross, which breeds in the Galapagos Islands, consists of many remarkable behavior patterns. The pair sway from side to side with necks extended forward, inspect the insides of each other's mouths, click their bills shut, wrestle with their bills, and point their heads downward, then up to the sky. What is the point of it all? *(Photos: Catie Rechten)*

8.2   Choosing a member of the right species

Matings between members of different species are very rare in nature. For example, when Monte Lloyd and Henry Dybas sampled 725 copulating pairs of two similar species of cicada (*Magicicada*) at a site near Chicago, they found only seven pairs containing members of two species; the other 718 (over 99% of the sample) were matings between males and females of the same species. Different species often do not get a chance to interbreed because they live in different geographic areas, or in different localities within an area (one species at the tops of trees, the other on the ground, for example), or they might be active at different times of day. Of the few species that do have the opportunity to interbreed, none do so frequently. They usually do not because one species does not respond to the sexual lures of the other.

It is adaptive for animals not to mate with members of other species. Hybrid offspring are usually inferior to those produced by matings of members of the same species, often because they are sterile. The best known example of a sterile hybrid is the mule. The mule is the offspring of a male donkey and a mare; if a female donkey mates with a stallion the result is a "hinney," which is also sterile. Natural selection will act to prevent animals from mating with members of other species if the offspring so produced would be sterile. Natural selection will favor animals that produce normal healthy offspring by choosing to mate with members of their own species, rather than producing sterile hybrid offspring. There have been many investigations of the factors animals use to ensure that they mate with members of the same species. Let us take crickets as an example. There are approximately 3000 species of cricket in the world. They do not all live in the same place, but in one habitat in North America there can be as many as 30 or 40 species breeding. The species do not interbreed because the males of each species sing a distinctive song (p. 48). Females are initially attracted by the song, as can be demonstrated by putting out a loudspeaker broadcasting the tape-recorded song of a male cricket: females of that species will approach it. Females, moreover, only approach loudspeakers playing songs of their own species. They are therefore using the song to choose a mate of the correct species. (Parasitic flies are also attracted to the loudspeaker. Parasitic flies are one of the hazards of the male cricket's sex life; they are attracted to calling males, on whom they lay their eggs. The growing parasitic larvae soon inactivate the male. It is a common risk in many species that by broadcasting to females, a male attracts predators and other enemies; some bats, for example, use the calls of male frogs to locate the frogs, which they then eat. Such was the fortune of the frog illustrated on the cover of this book.)

The importance of song in the species recognition of crickets can be shown by another experiment. Female crickets can be trained to walk along a Y-maze (Figure 8.2), on which they have to turn to the left or right. If, for instance, the loudspeaker on the left were playing the song of the female's own species, while the one on the right played the song of some other species, the female turns to the left. The location of the songs can be reversed to control for any bias the female may have for turning in one direction or the other.

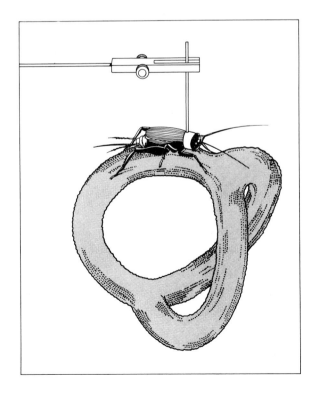

Figure 8.2
As the tethered female cricket walks along the Y-maze it moves beneath her feet. Loudspeakers play the songs of different cricket species from her left and right. Which song she prefers can be measured by which direction she turns on the maze. *(After Bentley and Hoy)*

The Y-maze has also been used to see how hybrid female crickets behave. If a male and female cricket of two closely related species are forced to mate, some hybrid offspring will be produced. When Ronald Hoy put these hybrid females on a Y-maze he found that they preferred the songs of hybrid males to those of either parental species. The result is interesting because it bears on the following problem. When a new cricket species evolves, the male song and the female receptor must change in a coordinated fashion: a change in one without a change in the other would be disastrous. The fact that the song and the receptor change together in the hybrids suggests that there is some control mechanism that can prevent the two from becoming uncoupled; such a mechanism would have the effect of making it more likely that, whenever a male or female cricket changed its song or song preference during evolution, other individuals of the opposite sex will have made the complementary change.

8.3   Sexual selection

Darwin put forward his theory of sexual selection to explain the bizarre, and apparently detrimental, sexual traits possessed by many species. He suggested that, in those species, males might compete among themselves to determine which males mate with the females, and the females might choose which males of the species they would mate with. If the males possessing the bizarre traits were favored by female choice, or in male competition, those traits could be maintained even if they decreased the viability of their bearers: the advantage in sexual success could compensate the disadvantage in viability. Both male competition and female choice are completely useless from the point of view of the total number of offspring produced: the females would get fertilized at all events. When sexual selection is operating, the females are mainly fertilized by particular favored kinds of males, rather than by a random selection of them.

A modern account of sexual selection would run something like this. We start from the gross difference in size of the male and female gamete. Each member of the next generation is formed by the fusion of one sperm cell with one egg cell, but the sperm cell is tiny—less than one millionth, often less than 100 millionth—the size of the egg cell. A male, therefore, feeding at the same rate and converting energy into gametes at the same rate as a female, can produce gametes at a much higher rate. A male is potentially capable of fertilizing hundreds of females, if he could only copulate with them. A male who did somehow manage to copulate with hundreds of females would be at an enormous advantage compared with a male who only copulated with one or two females. Natural selection will favor any adaptation in a male that enables him to copulate with more females. This is the theoretical basis of male competition. Females, by contrast, gain little or no advantage by mating with many males. Selection on females does not simply favor those who copulate with as many males as possible. The rate of reproduction by females is limited by the rate at which they can produce eggs, not by the rate at which they copulate. However, if males differ in their quality, it could pay females to be choosy about which males they will mate with. If one male were defending a better territory than another male, the female might be selected to mate with the better male. This is the theoretical basis of female choice. There is not normally any corresponding selection on males to be choosy about who they mate with. Males are selected to mate with as many females as possible. Any male who chose not to mate with certain kinds of females would mate less than his nondiscriminatory competitors.

Natural selection will therefore favor adaptations in males that enable them to mate with more females, but it will favor discrimination in females if males vary in their quality as mates. This general pattern follows from the cheaper cost of gamete production in males than females. Males can produce sperm at a faster rate than females can produce eggs, and are therefore selected to allocate more of their time to searching and competing for mates than are females. Because males spend less energy in manufacturing each gamete than do females, males can potentially produce more offspring, in a greater number of productive matings. Because males are able to reproduce more, by chance some indeed may do so. The variability in reproductive success of males will therefore be greater than that of females. These sexual differences may be illustrated by an experiment of A.J. Bateman on the fruit fly

Drosophila. He set up several experimental cages, each containing five males and five virgin females. Only 4% of the females did not mate, and even those were vigorously courted, but 21% of the males did not mate at all, even though they courted hard. The most successful males produced almost twice as many offspring as the most successful females (Figure 8.3). With sexual selection the success of different males varies highly. Some males are very successful, many are not. Females are less variable in the number of offspring they produce. The greater variability in the reproductive success of males could just be a random result: by chance some individuals will meet more mates than do others, and this will have a larger effect on the variability of reproduction of males, because in males more of the random encounters can result in mating. If a male meets a female he is unlikely to be prevented from mating merely because his sperm is in short supply (although this can happen); if a female meets a male she may well be short of an egg. This is a fundamental difference from which, in Darwin's theory, other sexual differences follow.

Figure 8.3
The relation between number of matings and reproductive success in experimental fruit fly (*Drosophila*) populations. The number of matings strongly influences the reproductive success of a male, but not of a female. *(After Bateman)*

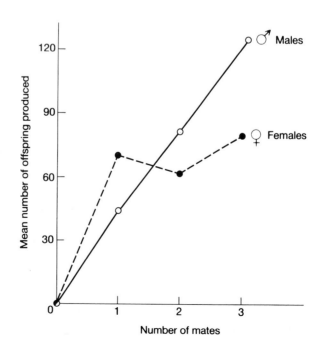

Chance alone is probably not the only factor influencing the relative variability in reproductive success of males and females. Female choice and male competition will also magnify the difference, because both processes increase the variability of male reproductive success. The effect of female choice and male competition will be to make some males even more successful than if those factors were not operating. It would take further work to determine the relative contributions of chance, male competition, and female choice, to the variabilities of reproductive success of male and female fruit flies in Bateman's experiment. But the fundamental theory predicts a strong relation between the amount of energy contributed to the production of the next generation by each sex, which sex chooses mates, and which sex fights or otherwise competes for mates. Let us now consider it in action with some real examples.

8.4 The struggle of males for females

Because the reproductive success of males is probably limited by the number of females they can attract and defend from other males rather than by their sperm supply, natural selection will favor any property in a male that enables him to mate with more females. This general principle finds its natural realization in the fascinating diversity of adaptations by which the males of different species seek to increase their share of the mating. Let us examine some examples.

The most obvious form of competition for females is straightforward fighting. We have considered the subject before, in the previous chapter, and we only need notice here the kinds of traits it leads to in males. It will favor strength, and fighting in its simplest form will favor increased size. In many species of toads (*Bufo*), for instance, the males cling to the backs of females for a few days before the female lays her eggs. Other males try to dislodge the sitting males from the females, by pulling them off. In experiments larger males are much better at dislodging smaller males from females than vice versa. There is an advantage in being large in males, and large males are more likely to mate with females than are small males. In other species males have evolved special weapons with which they fight over females. The narwhal's tusk is an example (Figure

Figure 8.4

The narwhal's spiral tusk, which is actually an extended tooth (up to 260 cm long), is mainly confined to males. The many functions that have been proposed for it range from sound transmission to drilling holes in ice, but it is really used in aggressive fighting among males. H.B. Silverman and M.J. Dunbar watched narwhals off Baffin Island and repeatedly saw males cross their tusks and strike them against each other. Males also often have body scars and broken tusks, suggestive of combat.

8.4). It is found only in males, who use their tusks to fight each other. Most males suffer severe wounds from these fights and in one sample from a narwhal population over 60% of the males had broken tusks. Other strange weaponry can be seen in the males of some kinds of beetles, and the antlers of deer have evolved for the same reason. In species that fight, a male can increase his effective strength by forming a coalition with another male, and ganging up on competitors. Male lions do just that. Groups of two or three male lions try to take over harems of females by forcibly evicting the incumbent males. Bigger coalitions of males are more successful in taking over harems.

However, physical fighting is not the only means by which males compete. When a coalition of male lions successfully takes over a harem of females, the first thing the males do is to kill all the young lion cubs in the pride. The cubs, of course, were fathered by the previous males, and are no loss to the new owners. There is even a gain to them. While a lioness is lactating for her cub she will not produce another cub, but after her cub is killed (or is weaned) she soon becomes ready to reproduce again. By killing the cubs, the males bring forward the time when they can start reproducing. Infanticide by males that have just taken over a harem is probably common in nature in many species; it has been observed several times in the Hanuman langur (*Presbytis entellus,* a species of primate) and has been anecdotally recorded in many other mammals.

A more subtle, less gory, form of male competition is the microscopic battle fought among the sperm of different males. In many species, if a female mates with two males, the second male by some means or other manages to fertilize many more than half the eggs. In the fruit fly *Drosophila,* for example, the second male fertilizes from 83% to 99% of the eggs. Males have evolved other kinds of counteradaptations to prevent sperm competition. Some male damselflies stay with their mates after copulation and fight off any other males who come near; only after the female has laid her eggs does the male leave her. Likewise, there is a parasitic acanthocephalan worm *Moniliformis moniliformis,* whose intermediate host is the cockroach and whose final host is the rat. The species has separate sexes. In the final reproductive phase, inside the rats, a male may cement up a female's genitalia after mating, probably to prevent her from copulating with further males. These male worms also sometimes cement up other males' genital openings, effectively castrating the victims. An extraordinary kind of sperm competition has been found in the hemipteran insect *Xylocoris maculipennis* by J. Carayon. Copulation in this species is achieved by injection; the male simply punctures the side of the female with his genitalia and squirts his sperm in. However, a male sometimes copulates with another male. His sperm then migrate to the victim's testes, and when the second male comes to copulate with a female, the first male's sperm will be injected into her.

8.5   Courtship and female choice

The classic example of a courtship sequence was worked out by Tinbergen, in the three-spined stickleback (*Gasterosteus aculeatus*). The male stickleback defends a territory (p. 226), and if a female wanders into it he usually attacks her. The male has a different appearance from the female: the male's belly is red, whereas that of the female is a glossy silver, distended (if she is ready to lay) with eggs. If a red-bellied stickleback stays in the male's territory he continues to attack it. But if a silvery, bulging-bellied stickleback stays he soon recognizes it as a female, and changes from attack to courtship. The first stage of the stickleback courtship is the male "zigzag" display. The male swims rapidly back and forth many times. The female, if receptive, responds to this by a "head up" display. This stimulates the male to lead her to his nest and show her the entrance by pointing his snout into it. The female may then enter the nest and lay her eggs (Figure 8.5).

Figure 8.5
Courtship usually consists of a recognizable sequence of male and female activities ending in mating. Courtship in the three-spined stickleback begins with a "zigzag" display by the male; if the female responds by showing a silvery egg-filled belly, the male leads her to his nest and shows her the entrance. The female may then enter the nest and lay her eggs while the male nuzzles her tail. After the female has laid her eggs and left the nest, the male follows her through and fertilizes the eggs. *(After Tinbergen)*

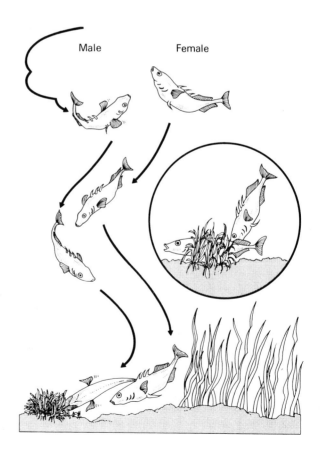

Why does courtship takes this form? Why all this swimming back and forth? Why not something else equally apparently arbitrary, such as blowing bubbles, or dropping pebbles? Tinbergen suggested that the zigzag display results from the conflict felt by the male between attacking the female and fleeing from her. It swims forward to attack, and away to flee (compare the argument about motivational conflict in the origin of autonomic signals, p. 182). Whatever its origins, the zigzag display is necessary if the female is to be attracted to the nest. Experiments have shown that males who perform the zigzag display at a higher rate (more swims back and forth per minute) are more likely to be successful at courtship: a female is more likely to mate with a male whose zigzagging is more energetic. In another experiment, by Semler, females were allowed to choose between mating with a normal red-bellied male or with a male of the ordinary nonbreeding stickleback green color. (In certain lakes in North America, there are male sticklebacks whose bellies do not turn red.) The females usually chose to spawn with the red-bellied males. Female choice, therefore, is part of the reason why such elaborate courtship displays are found in males. Males continue to court in the way they do because if a male were to stop producing part of the species' typical display, females would not mate with him. Male courtship behavior that does not increase the chance of mating is eliminated by natural selection.

Female choice is thought also to explain the evolution of bizarre, highly exaggerated displays, such as those of peacocks and other pheasants. The suggestion was first made by Darwin, but was made more explicit by R.A. Fisher. It is not enough only to point out that female choice might favor a bizarre male trait; if the trait is deleterious, one must also explain why the female preference is not then selected against, for the females will produce sons with the bizarre trait. Fisher imagined the following evolutionary sequence. Initially, males with slightly longer than average tails might have been fitter than average; females who preferred to mate with them would have an advantage because they would produce better sons. Because the preference is favored in females, it will become more common. Now the advantage to longer tailed males will increase, as they are both more fit and preferred by the females of the population. Under this twin pressure, tail length will increase; female preference and male trait will reinforce each other, and evolve together in a "runaway" process. Male tail length could now evolve to so great a size that it decreased the male's viability: after the preference has been established in the majority of the females in the population, it can balance a disadvantage in the preferred trait.

Fisher also suggested that the balance of advantage in mating and disadvantage in survival could be stable. Once the stage had been reached at which males possess traits that reduce their survival, we might think that natural selection would favor females who produced sons with shorter tails (or whatever the "runaway" trait is). But consider how selection would work on an individual female who did not preferentially mate with a male possessing the deleterious trait. She would indeed produce sons who would have higher survival than average. However, that advantage would be cancelled when they grew up. They would grow up in a population in which almost all the females prefer males with long tails, and they would therefore be rejected as mates and fail to reproduce. The female's mating preference would not spread. There is an advantage to females of sharing the majority preference, because then they produce sons who are likely to be successful. The theoretical details of the Fisher process are uncertain, and there is another theory of the evolution of exaggerated, deleterious male characters, a theory derived from the ideas we met in Chapter 7 (p. 185) about honest signaling. But let us now move on to a test of Darwin's general suggestion that female choice maintains exaggerated male traits.

If the theory is correct, the females in a species like peafowl must prefer those males with the most exaggerated tails. If that preference is not present, there would be no balancing force to counteract the advantage of smaller tail size, and the equilibrium should break down. The crucial test is on the preference; if it does not exist, the theory is wrong. Until a few years ago this test had not been performed in any species with a bizarre male form, and the theory remained a speculation, tentatively accepted in the absence of any plausible alternative. But in 1981 Malte Andersson tested for a preference in long-tailed widow birds (*Euplectes progne*), and provided the first strong evidence for Darwin's theory.

Long-tailed widow birds live in Kenya; the males have, among other sexual modifications, a far longer tail than the females.

Do females prefer to mate with males of longer tail length? Andersson tackled the question directly by experimentally altering the tail lengths of the males. The males fell into four categories: nine had their tails cut off and a longer one stuck on (with quick-acting glue), nine had a shorter one stuck on, and, as controls, nine were left intact, and nine had their own tail cut off and then immediately glued back again. Andersson measured the attractive power of the 36 males before and after the treatment and found that the number of nests on the territories of the artificially lengthened males was indeed higher (Figure 8.6). Actually, the

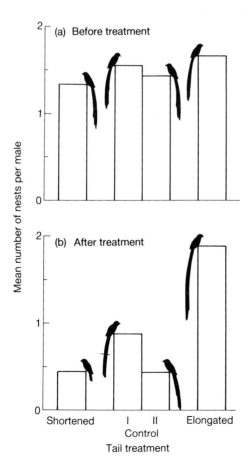

Figure 8.6
Female long-tailed widow birds prefer to mate with longer tailed males. (a) At the top are the mating successes of four groups of males before experimental treatment—they are similar. (b) The males then either had their tails shortened, lengthened, cut off and replaced (control I), or left alone (control II). The males with lengthened tails now enjoyed higher mating successes; the males with shorter tails were less successful. (*After Andersson*)

experiment is not perfect because, strictly speaking, Andersson only showed that the number of nests on a male's territory was related to tail length; the eggs could have been fertilized by another male, and the female then attracted to the territory of a male with a longer tail. However, in the absence of evidence, this is a pedantic objection. Andersson's experiment strongly supports Darwin's theory. Anders Pape Møller has subsequently obtained similar results in barn swallows (*Hirundo rustica*). Female choice is probably the reason why exaggerated sexual traits are present in the males of many species.

8.6   Systems of reproduction

The mating system of a species is the way in which the sexes associate for breeding. The two main properties of a mating system are the amount of time that the sexes associate for, and the number of males and females in a breeding group. In some species, the sexes do not associate at all: in sessile species, such as bivalve molluscs, and in some mobile species (such as some kinds of starfish) the sperm and eggs are simply discharged into the water, where they fuse. In other species, however, the sexes form a definite association for breeding. We shall be concerned here to establish what the main kinds are, and also to ask why they take the form, and occur in the species, that they do.

There are three main categories of breeding systems: monogamy, polygyny, and polyandry. (A fourth kind, called "promiscuity" is also recognized. In a promiscuous species, males and females form only short reproductive associations, and with many members of the other sex.) In monogamy, a single male breeds with a single female. Monogamy is rare in most animal groups, but is common among birds. Over 90% of bird species are monogamous. In a typical monogamous bird like the robin, the male and female stay more or less apart during the nonbreeding season, and only pair up shortly before breeding. They cooperate in preparing the nest, after which the female lays the eggs; both parents take turns in incubating the eggs and, after the eggs have hatched, they both bring food for the young. Similar breeding systems, with biparental care and monogamy, are used by some cichlid fish and several species of primates. Monogamy, but without parental care, is found in a few invertebrates: the limnorid isopods that bore into shipwood live in pairs, but the young look after themselves; likewise, some wood-boring scolytid beetles live in pairs but have no parental care in the ordinary sense of the word—although the young may benefit from the proximity of their parents.

In polygyny, in its purest form, a single male mates with several females. A minority of males accomplish most of the mating, and many males die without breeding. Polygyny, often combined with a degree of polyandry (as each female may mate with several males), is the most common breeding system found in most arthropods, fish, amphibians, reptiles and mammals. Polygynous systems can have long or short associations of the sexes. The association is long term in the species in which a male lives with a group of females. The wrasse *Labroides dimidiatus* is an example (Figure 8.7). Males of this fish species defend a group of three to six females, and forcibly prevent other males from mating with them. *Labroides dimidiatus* has the additional interesting habit of sex change. When the male of the group dies, the biggest female in the group quickly changes into a male and takes over the group.

The association between the sexes is very brief in another kind of polygyny, called a "lek" mating system; the bird species called sage grouse (Figure 8.8) and the ruff (*Philomachus pugnax*) are examples. The males defend special territories that are used only for mating. The male is briefly visited by a female for copulation, and the female then leaves and the male awaits his next visitor. There are relatively few territorial sites, which the strongest males occupy; these sites (or "leks") tend to remain in the same place every year. The few males holding territories on the lek do nearly all the mating. After a female has mated she departs from the lek to lay her eggs and rear her young by herself.

Polyandry is conceptually the opposite of polygyny, although in practice both systems can occur in the same population. In pure polyandry, each male mates with only one female, but a female (if successful) may mate with several males. It is much rarer than polygyny, but some examples do exist. The kind of bird called the jacana is one. The female jacana defends a group of males; she lays a clutch of eggs for each of her males,

Figure 8.7
The coral reef-inhabiting wrasse *Labroides dimidiatus* is a cleaner fish. (a) A wrasse emerging from the mouth of a grouper (*Plectro-tromus maculatus*), which it has been cleaning up. The grouper allows the wrasse to enter its mouth to take out skin parasites and any other edible items. The grouper then allows the wrasse to come out unharmed, and swims away. The grouper loses its parasites, and the wrasse gains a meal. The wrasses are territorial, and the same grouper may later return to the same cleaner wrasse. On the wrasse territory, a male defends a groups of females; when the male dies, one of the females changes sex and takes over the group. (b) Territorial encounter between males.
(Photos: D. Ross Robertson)

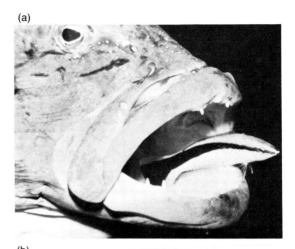

and it is the male who incubates the eggs and rears the young, unaided by the female. The female is bigger than the male, and forcibly prevents other females from mating with, or laying eggs for, any of her males.

Why are some species monogamous, others polygynous, and still others polyandrous? One important association is between the mating system and the mode of parental care. Even the few examples we have seen illustrate how species with biparental care tend to be monogamous, species in which only the female looks after the young tend to be polygynous, and species in which only the males care for the young tend to be polyandrous. The association, moreover, makes sense. Sexual selection only operates because individuals of one sex contribute more to the

Figure 8.8   Sage grouse mate on leks, in which males aggregate and defend territories from each other and are visited by females. The males are larger than the females and have modified tail feathers (left) and enlarged white esophageal sacs, which they puff up during the "strut" display (right). *(After photos by Haven Wiley)*

production of offspring than the other. Normally the productive sex is female. Males then compete, females choose, and the mating system is polygynous. But if the fundamental variable is altered, so too are the theoretical predictions. If both sexes work to produce offspring, as in a species with biparental care, the selection for males to compete for matings is relaxed, and monogamy may result. If males do the majority of the productive work, as in jacanas, the whole pattern of sexual selection may be reversed. The females evolve to be larger and brighter colored than the males and compete to control the domestic skills of males against other acquisitive and territorial females.

The association of parental care and mating system provides only a partial solution to the problem. Most species completely lack parental care, but they are by no means uniform in their mating systems. Many are polygynous, as we should expect, but others, like the wood-boring isopods and beetles, are monogamous. Moreover, the association of parental care and mating systems is imperfect. In many fish, like the three-spined stickleback, the males provide all the parental care, but the mating system is polygynous. Other factors must be at work.

However that may be, the association of mating system and parental care suggests another question—why do species differ in their sexual division of parental labor? The incidence of biparental care presents no great puzzle. It is presumably used by species in which it takes two adults to provide for the young. But what of species in which, it appears, only

Figure 8.9

(a) A male pycnogonid of the species *Boreonymphon robustum,* covered with his offspring, whom he is carrying. Most species of pycnogonids transport their offspring only during the egg stage, but this species carries its young too. Male pycnogonids have a special pair of ovigerous legs; they can be seen here at the front, ending in pincers. (b) Seahorses, such as this pregnant male *Hippocampus antiquorum,* are unique (with some pipefishes) in the animal kingdom in that the males are impregnated with eggs by the female, whose elongated ovipositor introduces the eggs to the male's brood pouch. *(After D'Arcy Thompson and Hesse-Doflein)*

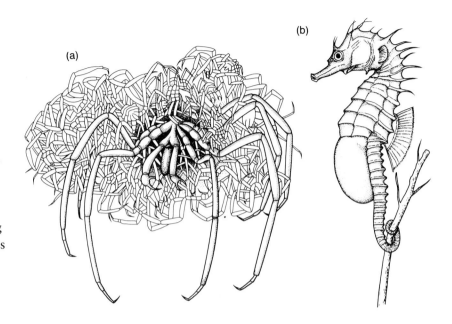

one parent is necessary? Why is it the male in some species but the female in others? Let us first examine some examples (Figure 8.9). Pycnogonids are small spiderlike creatures, ½–6 inches long depending on the species, which are not uncommon on seashores all over the world, clinging to sea anemones or the bottom of rocks. They have eight pairs of walking legs, with their genital openings situated in the top segment. At mating, in an operation never exactly described, the sexes are said to bring their genital openings into opposition and release their gametes. The male has an extra pair of legs at his front end, called ovigerous legs. He uses these to carry the female's eggs, in a ball, for a few weeks. During that time he may mate with other females, and add the eggs to his collection. I have seen male pycnogonids of the genus *Nymphon* on the shores of Britain carrying as many as six bundles of eggs, each from a different female.

The kind of paternal care found in some fish is more familiar. In the common river genus *Cottus,* in sticklebacks, and in seashore blennies, the male defends a territory; the female visits him and lays her eggs, which he then fertilizes; the female departs; and the male then looks after the eggs. Maternal care is used by many more groups: spiders, crabs, prawns, wasps, frogs, birds, mammals—no one has ever counted them all. They do, however, share one feature different from the paternally caring fish, and perhaps different from pycnogonids: they nearly all have internal fertilization.

There is indeed a suggestive correlation between the mode of fertilization and the sex that ends up looking after the young. Paternal care tends to be associated with external fertilization, maternal care with internal fertilization. The association is imperfect. A few years ago, I counted the number of families of animals with paternal care that had the two modes of fertilization: the result (now out of date) was 42 with external, to 20 with internal fertilization. No one has done the analogous count for all groups with maternal care. My result does show there are exceptions—the jacanas are one—but the rule seems to have enough generality to invite explanation. One possibility was put forward by Richard Dawkins and Tamsie Carlisle. They supposed that one sex was going to care for the eggs, but that selection on both sexes would make them, if possible, desert the eggs and thus force the other sex into doing the work. There would then be an evolutionary race among the sexes to run away first, and leave the other "holding the baby." The winner of such a race might well depend on the mode of fertilization. With internal fertilization in the female, the male could run away before the female could lay her eggs; whereas with external fertilization, if the eggs are released first, the female may be the sex with a head start. The theory is incomplete because, as we have seen, there are exceptions to the association it aims to explain. However, it may provide a part of a general theory, which might relate the mode of fertilization, the kind of parental care, and the system of mating, to explain the several general types of reproductive systems.

8.7   Summary

1. Courtship behavior enables an animal to find a partner of the right species; this is accomplished in crickets by acoustic signals.
2. The need to recognize mates of the right species cannot explain the bizarre development of sexual traits. Darwin proposed his theory of sexual selection to account for sex differences of structure and behavior. According to that theory, in most species, males compete for, and females select, mates.

3. Males in nature do compete for females in diverse ways, including obvious physical fighting and subtle invisible competition among sperm within the female reproductive system.

4. Female choice probably has selected for many of the properties of courtship, such as the "zigzag" display and red belly of male sticklebacks. Female choice is also the only known valid explanation of extravagant, and probably deleterious, male traits such as the peacock's tail.

5. Female choice has been experimentally tested in a species of widow bird in which the males possess long tails. Females prefer males with longer than average tails.

6. The sexes form various kinds of associations for breeding (called mating systems) in different species; some are monogamous, others polygynous, and others polyandrous.

7. The various mating systems can be partly explained by the mode of parental care; monogamous species often have biparental care, polygynous species maternal care, and polyandrous species paternal care.

8.8   Further reading

Trivers (1985) and R. Dawkins (1989) introduce the theory of sexual selection. Darwin (1871) remains an excellent review of the sexual traits of animals, with many thoughtful comments that are still relevant and are of more than merely historical interest. Bentley and Hoy (1974) describe their work on crickets; and Dethier (1992) describes the crickets of the northeast. Møller (1988, 1989, 1990) describes similar work to that of Andersson, but on barn swallows; Smith and Montgomorie (1991) confirmed Møller's result with North American barn swallows. On mating systems, see Krebs and Davies (1993), and Wiley (1991) on leks. Wingfield et al. (1990) have done interesting work on the hormonal associations of mating systems in birds. On parental care, see Clutton-Brock (1991).

CHAPTER 9

Conflict and Social Life

Fighting between animals can be unrestrained, but in many cases it is restrained. We begin by considering why restraint should have evolved in fighting. We then move on to look at dominance, in less and more socially complex species, and at territoriality.

9.1   Introduction

In nature, many animals are competing for limited resources. Animals could lead more comfortable lives if there were not so many competitors, each seeking more space to live in, more food, and the chance to produce more offspring. Animals passively compete for resources by taking as much as possible for themselves, but they will also, in some circumstances, actively fight for them (Figure 9.1). Fighting is the most overt, naked form of competition for resources.

We might expect that aggressive fighting would be common in nature. Natural selection will favor animals who compete successfully, direct fighting is the most obvious form of competition, and an animal is unlikely to increase its chance of success in a fight by pulling its punches. Darwin certainly thought that aggressive fighting was common. His work *The Descent of Man* (1871) contains several sections on "the law

Figure 9.1
Fights among males of the Central American lamellicorn beetle *Podischnus agenor* (a) usually take place in vertical tunnels within sugarcane. William Eberhard built artificial cane burrows (b) to watch fights. Fights take place when a male enters a tunnel where another male is resident. The aim of the fight is to push the opponent down the tunnel to its entrance, clamp him, swing him free, and then let him drop. The resident here has clamped his opponent and is lifting him away from the tunnel entrance. The invader is hanging onto the shredded sugarcane to try to prevent his fall. *(After Eberhard)*

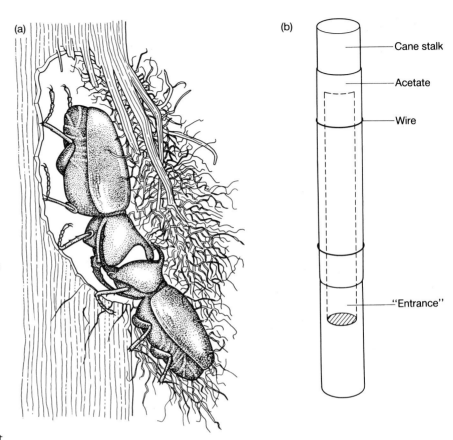

of battle" in different animal groups; he collected a large number of anecdotal observations, all of active fighting, blood-letting, and carnage. He does not hint at any restraint in animal fighting. This must be because he did not know any observations of it: Darwin was interested by every possible form of moral behavior in animals, and had he known of restrained fighting, he would have written about it.

The more detailed observations of animal behavior made in this century suggest a different conclusion. Active aggression is now thought to be exceptional. Animals apparently avoid unrestrained battles, and in this chapter we shall discuss three kinds of social behavior that reduce the amount of aggressive fighting. These facts of restrained fighting, however, in their turn pose something of a theoretical paradox. Darwin's own account of the unrestrained "law of battle" in animals readily fits in with the theory of natural selection; therefore, as his account of the law of battle is now thought to be wrong, we are left with the problem of reconciling new observation and established theory. We shall have to reconsider how Darwin's theory should be applied to animal fighting, to see whether natural selection might in some circumstances favor animals that avoid aggressive conflict.

9.2 Ritualized fighting

Most animal fights are restrained, or, as it was termed by early ethologists, "ritualized." (The word here has a different yet related meaning from that in the evolution of signals. The "ritualization" of a signal means the way a behavior pattern evolves from its initial nonsignaling state into a pure signal; the ritualization of fighting refers to the way fights are restrained and conventionalized, rather than all-out fights with tooth or claw. "Ritualized" fights often include "ritualized" signals.) Animals avoid using their most powerful weaponry when fighting other members of their species. Rattlesnakes are a clear example. A rattlesnake possesses a powerful poison, which it uses against prey and dangerous enemies. However, when fighting against another rattlesnake it does not use its poison fangs. Instead the two rattlesnakes fight in an energetic, conventional ceremony in which each tries to push the other to the ground. The loser, after being floored, retreats. The contestants come out of the fight relatively uninjured.

A ritualized fight will often proceed through several stages, like a tournament. According to an account by Konrad Lorenz, fights between cichlid fish of the species *Cichlasoma biocellatum* pass through up to three stages, at any of which one of the contestants may drop out. They start with broadside displays, move on to tail beating, and finish with harmless mouth wrestling. Each fish, according to Lorenz, moves on to the next stage only when the other is ready. The ritualized nature of the contest is made particularly clear by what happens if one of the fish finds itself with a temporary, but irregular, advantage. As Lorenz wrote:

One of them may be inclined to go on to mouth-pulling a few seconds before the other one. He now turns from his broadside position and thrusts with open jaws at his rival who, however, continues his broadside threatening, so that his unprotected flank is presented to the teeth of his enemy. But the aggressor never takes advantage of this; he always stops his thrust before his teeth have touched the skin of his adversary.

The fighting of male red deer, which has been studied by Tim Clutton-Brock and his colleagues on the island of Rhum off the west coast of Scotland, likewise passes through up to three main stages (Figure 9.2). They start by roaring at each other. The second stage is a kind of broadside display; it is called the "parallel walk," and in it the two males walk back and forth alongside each other. Only after that may they lock antlers and push each other. Red deer do actually inflict physical injuries in fighting. Minor cuts and bruises are common—some 25% of males are injured in this way each year—but serious injuries, such as a poked-out eye, have also been recorded. Six percent per year of the males watched on Rhum suffered permanent injury. Although the peacefulness of animals should not be exaggerated, even in red deer the level of aggression is considerably below the maximum possible. Less than a quarter of contests ever reach the final, antler-pushing stage, and even then the deer stag only pushes its opponent's antlers rather than trying to spike his softer flanks; antlers could perhaps be used more dangerously than they are.

Why do animals perform their restrained ritual contests rather than fighting all out straightaway? John Maynard Smith has suggested the following subtle answer. An animal that fought dangerously might well beat its restrained opponents. Natural selection would then favor the animals that fought dangerously, and they would increase in frequency in the population. But as the dangerous fighters increased in numbers they would increasingly have to fight against other dangerous fighters rather than against timid, restrained fighters. They would then be likely to injure each other badly. The risk of injury might become so great that the timid fighters, which never get hurt because they always run away

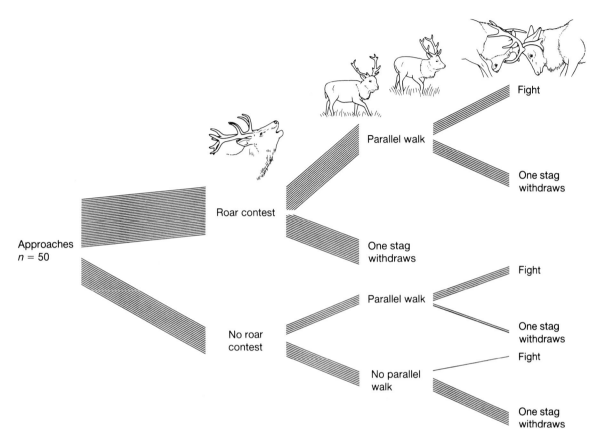

Figure 9.2   Fights of red deer stags pass through up to three main stages: roaring, parallel walks, and antler clashes. Fights may end at any stage with the retreat of one individual. *(After Clutton-Brock TH, Albon SD. The roaring of red deer and the evolution of honest advertisement. Behaviour 1979;69:145–170.)*

first, might then be at an advantage. The advantage to dangerous fighting is "frequency dependent." It is a good strategy when it is in low frequency, but not when it is common. The frequency dependence here is interesting, because although dangerous fighters always beat restrained fighters in a one-to-one encounter, still natural selection does not end up producing a population of dangerous fighters. It comes to a balanced equilibrium, with a proportion of dangerous and restrained fighters present. Maynard Smith's argument is sufficiently clear that it can be formulated mathematically, and the equilibrial balance calculated (Box 9.1). In

---

BOX 9.1   THE HAWK-DOVE GAME

Maynard Smith's argument applies game theory to animal behavior. In its simplest form, the argument can be reduced to a model with two strategies. (A "strategy" in game theory is the course of action an individual follows in some specified circumstance, such as a fight.) The two strategies are called "hawk" and "dove." The two words have nothing to do with the real bird species called hawks and doves; they are derived instead from military metaphor in humans. "Hawks" are animals who fight in an unrestrained aggressive manner; "doves" are retrained fighters, and they run away when they meet an opponent who fights for real. The particular model of hawks and doves is not supposed to represent the strategies of real animals; it is supposed to contain the essence of the biological problem, and allow us to see what natural selection will favor. In these terms, the question becomes that of why natural selection does not produce a population consisting entirely of hawk strategists. To answer it, we calculate the effect of the dangerous or restrained fighting strategy on the number of offspring an individual produces, and then work out which strategy (or combination of strategies) natural selection will favor at equilibrium.

1. Work out all the possible payoffs to each strategy.

Let us suppose that the animals contest a resource that is worth W points to the winner, and zero to the loser. Let us also suppose that serious injuries inflicted by hawks cost D points. (The points should be understood as the effect of a contest on the number of offspring left by the contestants. They are called "payoffs.") It is not straightforward to calculate how many points each strategy will recover from each kind of possible encounter. If a hawk meets a dove, the hawk wins W points and the dove

Box 9.1   Payoff matrix for the hawk-dove game. The payoffs are written for contestant A.

Contestant B

		Hawk	Dove
Contestant A	Hawk	$\dfrac{W-D}{2}$	W
	Dove	0	$\dfrac{W}{2}$

takes zero. If a dove meets a dove, neither will fight, but we might suppose that they settle the contest amicably in such a way that each has an equal chance of winning. Then, on average, if an individual fights many contests in a lifetime, a dove wins $W/2$ from its fights with other doves. And when a hawk meets a hawk? They will have an unrestrained fight. Let us suppose that one will suffer serious injury first and be the loser, and that on average a hawk has an equal chance of winning or losing. Then, from many enounters, the hawk's average payoff will be $(W-D)/2$. The four payoffs can be written in a payoff matrix (Figure 8b.1).

2. Work out which strategy is favored.

In a population mainly consisting of doves, the animals will nearly always play against doves, and "hawk" will be favored because its average payoff is higher in these fights: $W > W/2$. But now consider a population mainly consisting of hawks. Dove wins on average zero, and hawk $(W-D)/2$. Which strategy is favored depends on the relative values of W and D: if $W > D$, hawk is favored; if $W < D$, dove is favored. If $W > D$, hawk is always the best strategy and restrained fighting does not evolve. If $W < D$, hawk is best when dove is common, but dove is best when hawk is common. A population at equilibrium should have some of each; there will be some restrained fighting.

3. Work out the proportion of hawks and doves at equilibrium when W < D.

The calculation proceeds as follows. First, for algebraic convenience, let us call the four payoffs (see Figure 8B.1), a, b, c and d, where $(W - D)/2 = a$, $W = b$, $o = cro$ and $W/2 = d$. We know that there must be some proportion at which the two strategies, hawk and dove, have equal payoffs because each of them is selected for when it is rare but selected against when common. We shall call the proportion of hawks "p": p of the individuals in the population are hawks and $(1-p)$ are doves. At equilibrium, hawk and dove must be doing equally well. That is,

average payoff to hawk = average payoff to dove.

If the payoff to either strategy is higher than to the other, the strategy with the higher payoff increases in frequency until the payoffs become equal.

Now we calculate the average payoffs. When p of the individuals are hawks, an individual receives the payoff it expects against a hawk p times, and the payoff it expects against a dove $(1-p)$ times. Therefore,

average payoff to hawk $= pa + (1 - p)b$
average payoff to dove $= pc + (1 - p)d.$

At equilibrium, these are equal,

$$pa + (1 - p)b = pc + (1 - p)d,$$

and we solve the equation to find out the value of p. The solution is

$$p = \frac{d - b}{a-b-c+d}$$

which, if we substitute for a, b, c and d from the matrix (Figure 8B.1), reduces to $p = W/D$. If, for example, the gain from winning was half the cost of losing, we should expect the population to contain half hawks and half doves, and the proportions would vary if the relative values of W and D were something else.

4. Interpreting the model.

Maynard Smith calls the strategy that natural selection produces (that is, the strategy that exists at equilibrium) the "evolutionarily stable strategy" or ESS. When $W > D$ the ESS is hawk; when $W < D$, it is a proportion $W/D(=p)$ of hawks and $1-(W/D)$ of doves. The ESS with only one strategy (hawks) present is an example of a "pure" ESS and that containing a proportion of each is an example of a "mixed" ESS. It is theoretically possible for the mixed ESS to be realized in nature either by a population containing W/D hawk individuals and $1-(W/D)$ dove individuals, or by a population whose members all behaved as hawks in W/D of their contests and as doves in $1-(W/D)$ of them.

The sense in which a strategy is evolutionarily stable is that no other strategy (among those considered in the model) can do better than it. A population of animals will evolve to an ESS and then remain there: we should expect to see animals in nature behaving according to an ESS.

short, his explanation of restrained fighting is that in nature there is an equilibrium between the risks of injury if dangerous fighting becomes too common, and the advantage of winning fights if dangerous fighting is rare.

The frequency dependence in the advantage of dangerous fighting is probably not the only reason why natural fighting is restrained. Another, perhaps more important reason is that animals differ in their strength, and weaker animals will be selected to avoid fights with stronger adversaries. During a tournament like that of the red deer for instance, the opponents can probably size each other up to see how strong the other is. If an animal is to avoid fighting with another, stronger animal, it must first test how strong his opponent is. The early stages of the tournament may be restrained so that the animals can try their strengths without risking their lives. The roaring and parallel walks of the red deer may be safe trials of strength. We saw in the last chapter how signals may evolve to be honest indicators of strength, and roaring in red deer may be an example.

Natural selection does favor those animals that are most successful in the competition for resources, but that does not mean it favors unrestrained aggression. The most successful animals may avoid dangerous fights in order not to be injured by a stronger, or excessively carefree, opponent. The avoidance of conflicts with stronger opponents is the reason why dominance relations develop—they are the subject of our next section.

9.3   Dominance

9.3.1 Dominance and social conflict

Dominance is a common, but not universal, kind of relationship between the members of a group, in which some animals—the dominant ones—have priority over others—the subordinate ones. The priority concerns access to such desirable resources as food, places to sit, and mates. The Norwegian biologist Thorleif Schjelderup-Ebbe made the first important observations on dominance in the 1920s. He studied the hens of the common domestic chicken. Since then so many ethologists have studied these birds that more is known about dominance in hens than in any other species.

When an experimenter sets up a new group of hens, they first fight among themselves. Gradually the hens learn to recognize each other, thereby learning which are the stronger, and which the weaker, individuals. Each hen then learns to give way to stronger hens than herself; she learns not to get into fights she would probably lose. Dominant hens assert their priority by pecking subordinate hens: pecked subordinate hens move out of the way of the pecker, to allow the dominant hen access to the nesting site, or roosting site she was using, or the food she was about to eat. The exact form of the dominance relations within a group of hens depends on the size of the group. There is a simple linear hierarchy in groups of less than about ten hens. This means, in a group of say ten hens, that the "boss" hen is dominant to all the other nine hens, the second hen is subordinate to the boss hen but dominates the other eight, and so on down the hierarchy. In larger groups the hierarchy can become more complex. "Loops" may form, for example, in which one bird, A, dominates another, B, which dominates another, C, which itself dominates A.

It is advantageous for a hen to be dominant. Dominant hens take the pick of the food and the roost sites; dominant males also copulate more with females than do subordinate males. Natural selection must therefore be favoring the dominant animals. What kinds of animals become dominant? In hens the dominant birds are usually larger in size. They also tend to have more of the male sex hormone testosterone in their blood: a hen can even be made to ascend its dominance hierarchy by injecting some of this hormone into her blood. There is evidence, from other kinds of animals, of many other factors that affect dominance. Parasites, for instance, affect dominance, at least in mice. W.J. Freeland set up groups of three mice, the different mice having been injected with different quantities of parasites. The mouse with the fewest parasites usually became dominant. Dominance hierarchies may also be influenced by sex. In chickens, males are usually dominant to females, but if there are many males and females in the group, the males form a separate hierarchy above that of the females. In other species, the sexes may form separate hierarchies, or males usually dominate females, or (as in lemurs and hyenas, for example) females usually dominate males.

9.3.2 Dominance in primate societies

Many different kinds of social organizations have been found within primates (section 10.5), and each kind tends to have its own type of dominance system. Primate social relationships are strikingly complex, and a characteristic of primate dominance is the influence of alliances between individuals. Whereas in hens and mice dominance appears to be a one-on-one affair, in which the dominant individual displaces the subordinate individual, in many primates sets of individuals within a social group tend to form alliances during a conflict. The dominance status of an individual is then determined in the short term by which other individuals are nearby when it is challenged, and over the long term by which other individuals it usually forms alliances with.

One well-studied system is found in macaques (*Macaca*), and some other Old World monkeys, and we can consider it as an example. Macaque social systems are organized around what are called "matrilines." Male macaques usually emigrate from their natal groups when they reach sexual maturity, and join another group; females usually remain in their natal groups. The females in a group are therefore related by long-term bonds whereas adult males are more peripheral to the society. In conflicts, females are often supported by female relatives—that is, by other members of their "matriline"—and a whole social group may contain a number of coalitions of female relatives, with relatively stable dominance relations with respect to one another. The relative who is most likely to support an individual is her mother; a young female takes on the rank of her mother because the mother supports the daughter in any conflicts, and whoever the mother can displace the daughter therefore can too (by calling on her mother's support). Rank therefore is not determined by age. Rank is inherited from mother to daughter, down the matriline. One interesting consequence has been found in the sex ratio. Dominance is advantageous in various respects, and dominant individuals have higher reproductive success than subordinates. Because rank is inherited, dominant mothers tend to produce dominant daughters and subordinate mothers tend to produce subordinate daughters. A dominant macaque can therefore gain more by producing a daughter rather than a son than can a subordinate macaque: the dominant mother's daughter is more likely to be successful when she grows up. And in macaques, dominant individuals indeed tend preferentially to produce daughters (Table 9.1).

In a society where dominance is determined by alliances, we can distinguish between an individual's "basic" rank and its "dependent" rank. The distinction was first described for Japanese macaques (*Macaca fuscata*) by Kawai in 1958. Kawai threw sweet potatoes between pairs of juveniles and noticed that the one who got the food was the one whose mother was nearby. Basic rank means rank in one-on-one encounters; dependent rank means rank when allies are present. Since then a series of experiments on the distinction have been done by Bernard Chapais. In a typical experiment there might be three female macaques, *A*, *B*, and *C*, and their three daughters *a*, *b*, and *c*. If the three daughters are put together, they will have some observable (basic) dominance rank relations. For example, *b* might be the most subordinate. Then Chapais might introduce *B* to the group of three juveniles, and the rank of *b* would then rise. He might then remove *B* and introduce *C*: *b*'s rank would then fall, and *c*'s rise. These experiments clearly illustrate the

Table 9.1 Dominant rhesus macaque (*Macaca mulatta*) mothers produce more daughters. The numbers are for two colonies at Cambridge, U.K. (a) One colony, from 1960 to 1981. (b) A new colony. (*Data of Simpson and Gomendio*)

	(a) 1960–1981 colony Numbers of:		(b) New colony Numbers of:	
	Daughters	Sons	Daughters	Sons
Mother:				
High rank	38	15	23	8
Low rank	32	54	7	12

importance of matrilineal alliances in determining an individual's rank in macaques.

Within a social group, the existence of dominance relationships reduces the amount of overt aggression. The main reason is that weaker individuals come to learn that they are weak, and therefore avoid entering into fights that would be effortless to the stronger, and dangerous to themselves. However, additional mechanisms may be at work. In the pigtail macaque (*Macaca nemestrina*), the most dominant male reduces the amount of fighting within his troop by policing any fights that break out. For this reason, when M. Oswald and J. Erwin removed the dominant male from a troop the amount of fighting among the remaining pigtails went up (Figure 9.3); the effect is not simply due to the removal of an individual, because no consistent increase in aggression followed the removal of either a relatively dominant or a low-ranking female. The "policing" of conflicts, however, is a second-order effect. The establishment of a settled dominance hierarchy itself is the main limit on aggression.

9.3.3 Hormones, dominance, and stress in baboons

Olive baboons (*Papio anubis*) live in social groups that contain a number of adult males, a larger number of adult females, and their infants and

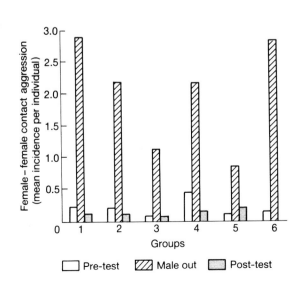

Figure 9.3
In each of six groups of pigtail macaques (*Macaca nemestrina*), when adult males were removed from the group the amount of aggression among the remaining females increased dramatically. When the males were returned the level of fighting went down again. (*After Oswald and Erwin*)

young. The males form a dominance hierarchy, and the females a relatively separate hierarchy; we shall concentrate here on some results for male dominance hierarchies in baboons, though the main interest of the findings is that they are probably more widely applicable. Dominant male baboons can displace subordinates from resting places, interrupt their mating attempts, take food out of their hands (perhaps after the subordinate baboon has laboriously dug it out or prepared it for eating), and expel them from shady bowers during the hot part of the day. In a study of stress in baboons at a study site in Kenya, Robert Sapolsky used observations of all those sorts of displacements to measure the dominance rank order of the males in the group; for simplicity, he then classified the top half of the ranks as "dominant" and those in the bottom half as "subordinate." He then measured the levels of certain hormones in these two classes of males, and studied how they changed during times of stress. We can focus on two hormones: testosterone (which is released by the testes) and cortisol (released by the adrenal glands).

For testosterone, the average basal level of the hormone was the same in dominant and subordinate baboons, but they responded differently to stress. Sapolsky cunningly exploited the stress that he as an experimenter inevitably imposed on the baboons to study their hormonal response to it: the stressful experience through which he measured hormonal levels was the occasion when the baboons were anesthetized, by shooting them with a blow-dart, to take blood samples. In both dominant and subordinate baboons, the testosterone level decreased after being shot, but although it immediately declined in the subordinates, in dominant baboons testosterone level initially increased and only declined later. The testosterone surge would give the dominant baboon an advantage in a real stressful situation, such as a physical challenge, because testosterone increases the supply of glucose to the muscles. The dominant baboon responds to stress by preparing to fight, whereas the subordinate does not.

Cortisol levels are good indicators both of social status and of how stressed an animal (including a human) is because cortisol levels are higher in subordinate animals and cortisol is released in response to stress. It has a number of consequences. Cortisol mobilizes the body's defenses for "fight or flight" and increases the supply of energy in the body. If this is merely a short-term response to temporary stress, it causes no problem, but in chronically stressed individuals, the overproduction of energy leads to hypertension and muscle wastage. Cortisol also suppresses the immune system, and Sapolsky duly found that subordinate baboons had lower quantities of lymphocytes in their blood as well as the preconditions likely to lead to atherosclerosis and increased risk of heart disease.

What is the mechanism causing high cortisol levels in subordinates? Cortisol is released as the end product of a chain of hormonal reactions (this is a common principle of many hormonal control systems—p. 72). Sapolsky traced the chain back and found that ultimate control of cortisol levels lies in the hypothalamus in the brain. The brain measures the level of cortisol in the blood and if the level is below what is appropriate, it stimulates the production of more; if the level is too high, it suppresses production. However, the measurement system in the brain of subordinate baboons seems to be insensitive to cortisol: they do not shut off cortisol production in the way dominant baboons do after a temporary surge. High cortisol levels are a consequence, not a cause, of subordinancy; a subordinate animal cannot be made dominant by lowering its cortisol levels. But the level of the hormone is the key to understanding the animal's physiological response to its social status.

Further work by Sapolsky has suggested that not all dominant baboons have low cortisol levels: some have high levels, like a subordinate. There may be two "styles" of dominance in male baboons. Some males are good at distinguishing real threats from nonthreats, given by rivals; they ignore the nonthreats and respond to the real threats by escalating the encounter into a fight. If they lose the fight, they "redirect" their aggression onto a subordinate baboon (Figure 9.4). Males that behave in this way have a "healthy" cortisol profile. Other dominant males show the opposite tendencies: they respond to nonthreats aggressively, are not particularly likely to fight when really threatened, and after they lose they rarely redirect their aggression on to a subordinate. These males have cortisol profiles like those of subordinate baboons. It appears,

Figure 9.4

Cortisol level comparison for male baboons showing two "styles" of dominance. The first pair of histograms, for example, compares the cortisol levels of those baboons who are most likely to distinguish a real threat from a nonthreat with those baboons who are least likely to. Dominant males, who distinguish real threats, respond to them by fighting, and redirect their aggression after defeat, have healthier, lower cortisol levels. *(Modified after Sapolsky)*

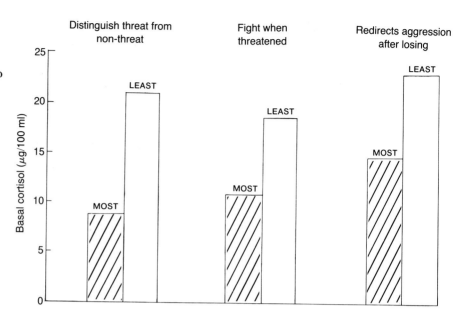

therefore, that the healthy way of dealing with the stresses of baboon society resembles what is suggested for handling stress in humans. "Stress managers" suggest that the ability to predict correctly and control social interactions, and to find harmless outlets for tensions when they arise, can reduce the deleterious effects of stress. Sapolsky's investigation of baboons may illustrate a general principle.

9.4   Territoriality

The three-spined stickleback (*Gasterosteus aculeatus*) is a fish about 2–4 inches in length that inhabits brackish and freshwater habitats in Europe and North America. It is a favorite animal for experiments on animal behavior. Sticklebacks swim around in schools in the winter; but in the spring, when temperatures increase and the food supply improves, male sticklebacks start to develop red bellies and become more aggressive. They each start to defend an exclusive area from other sticklebacks. This defended area is the male stickleback's territory. If another stickleback swims into a male's territory, the owner swims hard at the intruder, and drives it off. But he will only attack the other fish if it trespasses onto his territory: if the first fish trespasses onto the other male's territory the roles are reversed. Niko Tinbergen demonstrated the site dependence of which fish attacks and which one flees in a simple experiment (Figure 9.5). He let two male sticklebacks, *a* and *b*, set up their territories in an aquarium. He then put *a* in one test tube and *b* in another. When he put both test tubes in *b*'s territory, *b* tried to attack *a* through the glass and *a* tried to flee; Tinbergen then moved both test tubes across to *a*'s territory and now *a* tried to attack, and *b* to flee. A territory, then, is an area that its owner tries to defend and from which intruders usually flee. It is a dominance relationship that depends on the individuals' position in space.

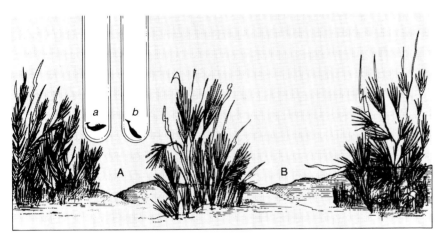

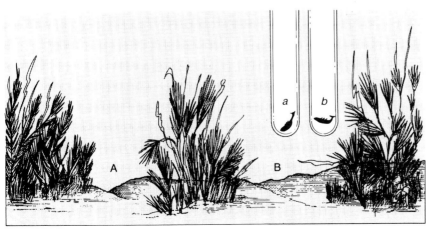

Figure 9.5

Which of a pair of male sticklebacks wins in a territorial dispute depends on where the fight takes place, as a simple experiment by Niko Tinbergen demonstrates. There are two fish, called *a* and *b*, with neighboring territories. Tinbergen put each fish in a glass tube. When he put the two tubes in *a*'s territory (A), *a* tried to attack and *b* to flee, and vice versa. Who wins depends on place, not relative strength. (*After Tinbergen*)

In contests between the territory owner and an intruder, the owner usually wins. Why should this be? One reason is that owners are often stronger than intruders, which are animals too weak to have been able to set up a territory. That cannot be the reason in Tinbergen's experiment, however; some other explanation must be found. The territory may be worth more to the owner than to the intruder, and the owner is therefore prepared to fight harder for it. The owner will have spent time finding out about his territory: where the best sites are for hiding, where the best food is to be found. The intruder lacks this knowledge; the territory would, to begin with at least, be worth less to him. A third possible reason, and one suggested by Tinbergen's experiment, is that a convention of "owner always wins" is used to settle territorial disputes.

The defense of territories is a widespread habit among many kinds of animals. In species in which the habit is found, it has often been shown that individuals without territories have lower survival or lower reproductive success than individuals who own territories. Different species have probably evolved territorially for different reasons; some functions are more common than others, and two of the main ones are feeding and reproduction. The stickleback's territory is for reproduction. Natural selection favors the habit in stickleback males because female sticklebacks will only mate with territorial males. After mating, the territory is advantageous in protecting the eggs and young fish. In other species, territories are defended for energetic, rather than reproductive, reasons. Let us consider two examples from birds.

The golden-winged sunbird (*Nectarinia reichenowli*) defends a territory of about 1000–2500 flowers for purposes of feeding; it feeds on the nectar provided by the flowers. The sunbird's territory is advantageous not only for the food it contains, but also because the food resources can be exploited more efficiently. F.B. Gill and L.L. Wolf, who studied the sunbird near Lake Naivasha in Kenya, measured the nectar levels in defended and in undefended flowers. There was on average more nectar in the defended flowers. This is because, after a sunbird has sucked out the nectar from a flower, the flower takes some time to replenish its nectaries. A territorial sunbird can time its visits to a particular flower such that its nectar has built up to a high level. Undefended flowers cannot be farmed in this manner, because another bird could come and drink from the flower during the interval before it has restored its nectar to the higher level. Defended flowers can therefore be exploited more efficiently.

If a bird is defending a territory in order to exploit the food resources on it as efficiently as possible, the bird should ensure that it defends a territory no larger than necessary. It takes energy to defend a territory from intrusive neighboring birds, and the larger the territory the larger the border that has to be guarded. It might be that if a bird tried to defend too large a territory the amount of energy it would gain from its territorial habit would actually decrease, as it spent more energy on extra defense than it gained from extra food supplies. This idea may be illustrated in a study of the rufous hummingbird (*Selasphorus rufus*) by Lynn Carpenter, D.C. Paton, and M.A. Hixon. This hummingbird migrates southwards down the west coast of the United States during late July and August. The flight burns up energy, and the hummingbird has to stop from time to time during its journey, to defend a territory and refuel. Carpenter and her colleagues watched the hummingbirds in the mountains of California. There, individual hummingbirds, whose weight has decreased to 3–3.5 g, stop for periods of a week or two; after they have put on about 1.5–2 g weight they then move on again. The rufous hummingbird, like the golden-winged sunbird, feeds on nectar, which it takes from a territory of about 60–4000 flowers. Carpenter kept track of the birds' weight changes by means of perches attached to spring or electronic balances (Figure 9.6). They observed that birds start to put on weight soon after establishing a territory (Figure 9.7a). But it is the relation of the rate of weight gain and the size of the territory that is more relevant here (Figure 9.7b). The rate of weight increase is lower if the territory is too large or too small, relative to an intermediate optimum. The points of Figure 9.7b are for successive days, and it appears that the bird adjusted its territory size to maximize its rate of weight gain.

Hummingbirds and sunbirds, then, defend territories in order to feed more efficiently on their defended flowers. The purpose of the stickleback's territory is different. In general, territoriality (like group

Figure 9.6
A territorial rufous hummingbird weighs itself by perching on a spring balance. *(Photo: Mark A. Hixon)*

living, as we saw in Chapter 6) has different functions in different animals. In all cases, however, the space will be defended to secure access to some limited resources. And despite the universal competition of animals for limited resources, settled territoriality will, like ritualized fighting and dominance, have the consequence of reducing the amount of naked aggression that can be seen in nature.

9.5   Conflict between groups

So far we have mainly been considering conflict between individuals. However, many animals live in social groups, and in these species there can be conflict between different groups of animals of the same species. The groups may each defend territories, and conflict may then be over space; in other cases, groups may fight over resources such as food. In

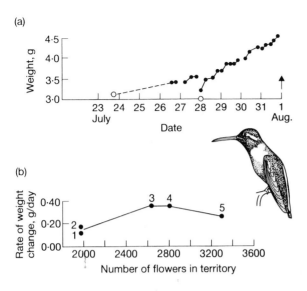

(a)

(b)

Figure 9.7

(a) An individual
rufous hummingbird
steadily increases in
weight during a week
of defending a
territory. On 1
August the bird flew
on. ○ Weights
measured by netting
the bird, ● perch-
balance weights of
free bird on its
territory. (b) The
weight gain of the
bird per day is
related to the number
of flowers on its
territory (territories
change size between
days). The five points
are for five
consecutive days, and
suggest that the bird
has an intermediate
optimum territory
size. Both graphs are
for the same
individual; eight birds
all gave similar
results. (*After
Carpenter, Paton and
Hixon*)

several species of ants a group from one colony may "raid" larvae from another colony and bring them back to their own nest, where they act as "slaves" after they have emerged as adults. A common observation in all these cases is that larger groups tend to defeat smaller groups when they come into conflict; Figure 9.8 shows two examples, from the capuchin monkey (*Cebus apella*), a monkey that lives in Central America, and acorn woodpeckers (*Melanerpes formicivorus*), a group-living woodpecker that is well known in the western United States for the damage it can do to all wooden structures, including homes and telegraph poles.

The advantage of larger groups in conflicts is one of the fundamental selective forces causing animals to evolve to live in groups. It is not the only such force, as we saw some other advantages in Chapter 6 and we shall meet another in the next chapter. However, the universality of competition for resources, and the clear advantage of larger groups in direct conflicts, means that in species of mobile animals—particularly species that exploit resources that (once obtained) can be used by more than one individual—conflict predisposes them to evolve cooperative, social ways of life.

Figure 9.8

Larger groups defeat smaller groups when groups of animals come into conflict: (a) acorn woodpeckers, (b) capuchin monkeys. The two graphs illustrate two ways of plotting the result. For the acorn woodpeckers, each point is for one victorious group, and shows the size of that group against the average size of the groups it defeated. That larger groups usually win is shown by the fact that 10/12 points are above the 45° line. For the capuchin monkeys, the percentage of conflicts won by a group is shown for groups of various sizes. *(After Harcourt, using data of Hannon et al. and of Robinson)*

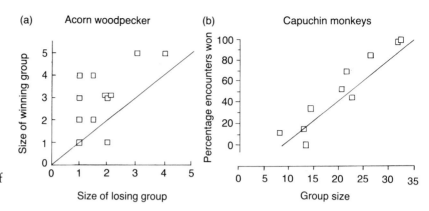

9.6   Summary

1.   Animals compete for limited resources, but the competition is often not an unrestrained battle to the death. In some species, contests can take the form of ritualized tournaments.

2.   Fighting may be restrained because the advantage of dangerous tactics will decrease as the habit becomes more frequent in the population. If the costs of injury are great, the advantages of aggression will eventually be limited by the risks of injury.

3.   Animals also restrain their aggression because it is disadvantageous to fight stronger opponents—it is better to run away. Thus "dominance" relations arise in many social species.

4.   Many factors influence the dominance of an individual—its size, strength, sex, health (in mice, at least), and age (in primates, but not in hens).

5.   In social groups, an individual's dominance can depend on which other individuals it forms alliances with. An individual has both a "basic" and a "dependent" rank.

6.   Levels of the hormone cortisol are higher in subordinate individuals. In male baboons, some dominant individuals also have high cortisol levels, and the cortisol level depends on the "style" of dominance that an individual displays.

7.   An individual (or a group) may defend a space, called a territory, from other individuals. When territories are formed, the dominance

relations of individuals depend on space. One stickleback will put another to flight when it is an owner attacking an intruder, but will itself flee from the same individual when it is an intruder on the other's territory.

8. Territoriality, dominance, and ritualized fighting all result in the fact that the amount of naked aggression to be seen in nature is less than the maximum possible.

9. In conflicts between groups, a group containing more individuals is usually more likely to win.

9.7   Further reading

Lorenz (1966) popularized the concept of restrained fighting, together with his own explanation of it, which is no longer accepted; see Dawkins (1989) and Krebs and Davies (1993) for the game theoretic explanation. Clutton-Brock et al. (1979) and Clutton-Brock and Albon (1979) describe the contests of red deer. On dominance in primates, see various authors in De Waal and Harcourt (1992), and Sapolsky (1990, 1994), Ray and Sapolsky (1992), and Keverne (1992) for the endocrinology. Tinbergen (1953) describes his experiment on sticklebacks, and Harcourt (1992) reviews the advantage of larger groups in contests between groups.

CHAPTER 10

Cooperation and Social

Life

Altruistic behavior requires special conditions if natural selection is to favor it, and we begin by establishing what those conditions are. We then see how to test whether the conditions apply in a real case, and use "helpers at the nest" as an example. We consider the social insects as the most complex cooperative societies ever to have evolved; the question of how animals recognize kin; and we use cooperative behavior in primates to illustrate a second theory of the evolution of altruism: reciprocity. We finish by looking at how parasites can manipulate other animals into behaving altruistically to them.

10.1   The natural selection of altruism

Altruism means the transfer of some benefit from the altruist to the recipient, at a cost to the altruist. It is characteristic of what we think of as the most highly developed forms of social behavior. This meaning of the term has an important difference from that of ordinary human conversation. When we call a human act "altruistic" we not only mean that the altruist has given something away, but also that the altruist intended, roughly speaking, to be kind; the word refers to subjective intent as well as the objective transfer of goods. Students of animal behavior, however, almost invariably ignore subjective intentions; they concentrate exclusively on observable units of behavior. In borrowing the word "altruism," therefore, they have altered its meaning to fit their scientific method. The word "selfish" has likewise been borrowed and stripped of its subjective connotation: behavior is "selfish" if it confers a benefit on the selfish actor, at some cost to the victim. In these terms, it is altruistic to give another individual a meal, and selfish to take one away, regardless of what subjective intent may lie behind the deeds.

Altruism interests biologists because, at first sight, it seems to contradict the theory of natural selection. Natural selection favors traits that increase the reproductive success (or fitness) of their bearers. Altruistic traits, however, must do the opposite. Indeed, they decrease reproductive success *by definition,* because, although I previously left the measurement of benefits and costs vague, the only universal currency in which to measure them is the currency of offspring. We may define an altruistic act as one that increases the reproduction of the recipient and decreases that of the altruist; a selfish act is the opposite, one that increases the reproduction of the selfish type and decreases that of the victim. With those definitions, natural selection ought only to favor selfish behavior. Altruistic behavior, however, that fits the definition, does exist. The sterile worker castes of social insects, who work to increase the reproduction of another individual and do not themselves reproduce, are one example, and we shall consider them and some other cases below. But let us consider the theoretical question first. How could natural selection favor sterility?

One possible answer is the theory of "group selection." So far we have discussed natural selection as if it favored behavior that increased the reproduction of the individual showing the behavior. According to the theory of group selection, however, it favors the reproduction of the whole group that the individual is a member of. Then an individual might sacrifice itself for the good of its group. We can imagine some groups made up of altruists, and others of selfish members. The altruistic groups would probably produce more offspring as a whole, and if natural selection does indeed work on the reproduction of whole groups, the greater reproduction of altruistic groups would be the reason why altruism evolved. However, it is a matter of controversy whether natural selection can favor a trait that increases the group's reproduction but decreases the individual's.

Consider what is going on within each group. If there is an altruist or two within a selfish group they will leave less offspring than average and the habit will be selected against. Vice versa, selfish individuals within a mainly altruistic group will take advantage of the altruistic deeds performed to them and (by definition) outreproduce the altruists. Within each kind of group, selfish animals will increase in frequency. Selection on individuals within a group and selection of whole groups are therefore favoring opposite traits: group selection favors altruism and individual selection favors selfishness. Which will win? Most biologists who have considered the question in detail have answered that individual selection is more powerful. The general reason follows from the different rates of the two processes: selection between individuals is faster than selection between groups. Individual selection within a group adjusts the relative frequencies of selfish and altruistic types every generation; group selection can only operate when a group goes extinct. The extinction of groups cannot take place faster than the death of their members, and in nature probably takes place much more rarely. The plodding process of group selection may favor altruistic groups from time to time, but individual selection will quickly reconvert each group to selfishness long before the groups go extinct again. At any one time, most animals' behavior will be selfish. Theoretical models in which group selection wins out require unrealistically high rates of group extinction, or unrealistically low rates at which selfish individuals can appear within pure groups of altruists. (An analogous argument has been used previously, for the case of cultural evolution and genetic evolution by individual selection, on page 115.)

If group selection is ruled out, altruism is a stronger paradox than ever. Altruism is exactly what individual selection should act to prevent. The most plausible solution to the paradox has been suggested by W.D. Hamilton. It comes from considering how natural selection will operate when interacting individuals are genetically related. Natural selection can work only on a trait that is heritable and it is easy to see how, in some circumstances, a heritable disposition to provide parental care might be favored. The caring male stickleback, for instance, fans his eggs in their nest. If the male is taken away, the oxygen concentration in the nest declines, and many of the eggs catch fungal diseases and die; parental care therefore is a benefit to the stickleback's offspring. It is also presumably provided at some cost to the male, for fanning is a vigorous activity. A heritable tendency to provide parental care, present in the male, would be inherited by the very offspring whose survival chances the caring habit was increasing, and thus would increase in frequency. When the heritable tendency first arose, it would be as a single genetic mutation. Because the mate of the individual with the mutant would not have it, by the ordinary pattern of Mendelian inheritance (p. 79) it would be passed on to only half the offspring (Figure 10.1). But the parent dispenses care to all its offspring, including the half that lack the gene for parental care. In order for the mutation to increase in frequency this dilution must be more than made up for by the benefits of parental care. Let us symbolize the increase in survival conferred by care with B (for benefit) and the cost to the male of his exertions C. If the mutant is to increase in frequency, $\frac{1}{2}(B)$ must exceed C (the $\frac{1}{2}$ being because only half the beneficiaries actually possess the gene). The figure of $\frac{1}{2}$ is technically called the "relatedness." Relatedness is the chance that a gene in one (specified) individual is also in another (specified) kind of individ-

Figure 10.1
When a mutant gene (M) arises it has a half chance of being passed on to an offspring. Offspring can be of four kinds, according to the combination of chromosomes they inherit from their parents, and the four kinds have equal chances of being produced. Two of the four kinds bear the mutant gene. The chance of any particular offspring bearing the mutant gene is therefore $\frac{2}{4}$, or $\frac{1}{2}$.

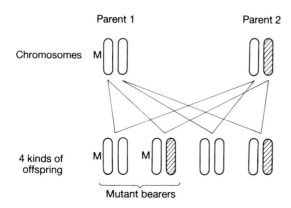

ual; it is written algebraically as r, and the formula for the condition in which the gene spreads is then $rB > C$. The formula is called "Hamilton's rule."

Parental care fits the definition of altruism. The key to an understanding of Hamilton's theory is that a genetically equivalent process can operate between other classes of genetic relatives. Let us see what the relatedness (r) is between a full brother and sister (Figure 10.2). Again, we are interested in whether a mutation will spread in the population from an initially rare state. We therefore suppose that it will be in only one copy in any individual. Suppose a female possesses a mutation: what is the chance it is also in her brother? The calculation is this. She has only one copy, which came either from her father (chance ½) or from her mother (chance ½). The chance that a gene in a parent is inherited by its offspring is ½; therefore, the total chance of the gene being shared between brother and sister through their father is ½ × ½ and through their mother is also ½ × ½. The total chance of sharing the gene is the sum of these two probabilities, ¼ + ¼ = ½. The relatedness between full siblings is one half. Similar arguments, of increasing complexity, can be made for all classes of relatives (Table 10.1).

Just as a mutation causing parental care will increase in frequency if the care is of sufficient benefit relative to its cost, so can a mutation causing altruism among any class of relatives. Take the case of full

Figure 10.2
Calculation of relatedness from sister to brother under normal diploid Mendelian inheritance. Consider a new mutant gene (∗) in a female. What is the chance that it is in her brother? The gene in the female has a ½ chance of being in her father, and a ½ chance of being in her mother. If it is in her father it in turn has a ½ chance of being passed on to her brother, giving a total chance that the gene is shared through her father of ¼. The chance of sharing the gene through her mother is likewise ¼. The total chance of sharing the gene is the sum of the maternal and paternal probabilities, ½. This is the relatedness of a sister to her brother (see Table 10.1).

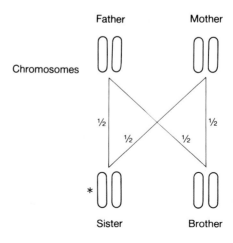

Table 10.1 Relatedness (*r*) among several different classes of relatives.

Classes of relatives	Relatedness
• Parent-offspring	1/2
• Full siblings	1/2
• Half siblings	1/4
• Grandparent-grandchild	1/4
• Uncle/aunt-nephew/niece	1/4
• First cousins	1/8
• Second cousins	1/32

siblings. Suppose that a mutation arises such that mutant individuals direct altruism to their siblings. The mutation, on average, will be present in half the recipients. If it is to spread, the extra reproduction (*B*) of those individuals must more than make up for the decreased reproduction (*C*) of the altruist. Again, therefore, $\frac{1}{2} B$ must exceed *C*. The appropriate figure for relatedness from Table 10.1 can be substituted to give the condition for altruism to be favored among any class of relatives. Altruism among uncles and their nephews will be favored provided $B > 4C$. The formula can be easily extended for mixtures of relatives. An act that benefits a group made up of 30% nephews and nieces and 70% offspring will be favored if $((\frac{3}{10} \times \frac{1}{4}) + (\frac{7}{10} \times \frac{1}{2})) B > C$, and so on.

Hamilton's theory of altruism makes a clear prediction. Animals should only be altruistic toward genetic relatives, and under quantitatively specifiable conditions. If that prediction is satisfied in real cases of altruistic behavior, the puzzle of altruism may be removed. If it is not, we shall need another theory. Let us consider two examples where Hamilton's theory seem to apply, and then two others where other processes are at work.

10.2 Helpers at the nest

Many species of birds are monogamous, and the usual breeding unit is a single male and female, who breed and care for their young together. In some species, the breeding pair are assisted by a number of other, nonbreeding individuals, called "helpers." Jerram Brown has defined a helper as "an individual that performs parent-like behavior toward young that are not genetically its own offspring." The actual work done by the helpers varies from case to case, but they may bring food to the nest and defend the young against nest predators. In many cases, the helpers are the progeny of a previous clutch of the adult pair they are helping: instead of departing to set up a nest on their own, they stay at the natal nest and help to rear their younger brothers and sisters. Helping is an example of altruistic behavior, and we can see whether Hamilton's theory applies to it.

The Florida scrub jay (*Aphelocoma coerulescens*) is a thoroughly studied species with helpers at the nest. The scrub jay is distributed widely across the western United States, and also has an isolated population that breeds in the shrinking areas of oak scrub in central Florida. A breeding pair of Florida scrub jays may be helped by up to six helpers, and the majority of helpers are either full or half siblings of the young they are helping. The crude prediction of Hamilton's theory is therefore correct: the altruistic helpers are genetic relatives of the recipients of their altruism. Moreover, among nonbreeding birds at a nest, the full siblings are more likely to help actively than are half siblings, and half siblings are more likely to help than are unrelated birds in the few cases in which unrelated birds attend a nest (Figure 10.3). However, we can also attempt a more exact test, because the species has been continuously studied in Florida since 1969 by Glen Woolfenden and his colleagues. The family trees of the birds in the population are known, and reveal whether individual helpers are full siblings, half siblings, or unrelated to the birds they are helping. (The pedigrees are reliable guides to relatedness in this species because genetic fingerprinting has revealed no evidence of extrapair fertilization.) If Hamilton's theory applies, the condition $rB > C$ should be satisfied (where B is the average benefit and C the cost of the altruistic act). We know r from the family trees, but how can we measure B and C?

"Benefit" and "cost" properly refer to the change in the lifetime reproductive successes of altruist and recipient, relative to the act not having

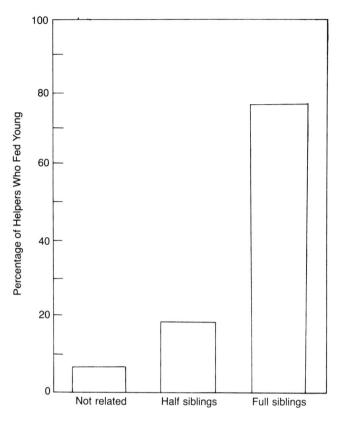

Figure 10.3
The nest of a breeding pair of Florida scrub jays may have full siblings or half siblings of the offspring attending it, or unrelated birds. A higher percentage of the full siblings actually help the young than do the half siblings, and a higher percentage of half siblings help than of attending unrelated birds.
(After Mumme)

been performed. The actual *B* and *C* therefore are unmeasurable, because they refer to a situation that does not exist (namely, if the act had not been performed). However, they can be estimated. Ron Mumme experimentally removed the helpers from 14 nests in 1987 and 1988, and measured the reproductive success both in these experimental nests and in 21 untreated control nests in the same area. The removal of the helpers significantly reduced the survival of the offspring (Table 10.2). Mumme found that the contribution of helpers is mainly in defending the nest against nest predators such as snakes and other birds. A nest with a helper is more likely to have a "sentinel" bird present at the nest at any time than is a nest without helpers, and nests with helpers can "mob" predators more effectively. (Mobbing is a kind of group defense of birds

Table 10.2 The survival of the young at nests of the Florida scrub jay, in nests at which the helpers had either been experimentally removed or left undisturbed. The offspring in experimental groups have lower survival during and immediately after the period of parental care, but not at the egg stage. These results are for 1987; the postfledging difference was similar in 1988, but the prefledging difference was not. (*After Mumme*)

	Experimental groups (helpers removed)	*Control groups (helpers present)*
Initial sample size	45	63
% survival from egg to hatching	67	68
% survival from hatch-ing to fledging	30	63
% survival to day 60 after fledging	33	81
% survival from egg to day 60	7	35

against predators, in which the birds dive at and harrass the predator. It is most commonly seen when birds such as crows mob domestic cats.) The young in nests with helpers are also fed more and (probably in consequence) survive better after fledging. Mumme's result enables us to make a rough estimate of the benefit (*B*) of helping as the difference between the survival rate of young in nests with and without helpers. In Table 10.2, the survival rate is increased somewhere between two- and five-fold if helpers are present. Let us use the lower estimate to suggest a minimum value for the benefit of helping (*B*). The survival of an average young scrub jay from hatching to fledging is increased from about 30% to about 60% if helpers are present: the difference is $63 - 30 \approx 30\% = B$.

The cost of helping is more difficult to estimate. It is the reproductive success a helper would have had if it had not helped. We can make an upper- and lower-bound estimate. The lower-bound estimate is zero if the helper had been unable to breed independently. This may be close to the true value of C in the saturated habitat of Florida scrub jays. It is thought that one of the main advantages of staying at the parental nest is the chance either of inheriting the territory or "budding off" another territory at the edge of it: most new territories are formed in this way. A young jay may have to stay at home in order ever to be able to breed independently. An alternative, upper-bound estimate of the cost is the reproductive success (30% per egg to fledging) of pairs without helpers. The justification of this estimate is that if the helper had bred by itself it would lack helpers (which are derived from earlier clutches) and thus achieve the success of an unhelped pair.

To apply Hamilton's formula in this case we have to notice that the helper's choice is between producing siblings and producing its own offspring. It should help its siblings if

$$r_{sib}B > r_{off}C,$$

where r_{sib} is the relatedness to a sibling and r_{off} is the relatedness to its own offspring, both of which are ½ for full siblings. (The small difference from Hamilton's formula as it was given before arises because there we imagined an individual either helping another individual or helping itself. Here the choice is between helping two kinds of other individuals.)

For the two estimates of cost, again staying with the figures for survival to fledging, the inequalities are approximately:

lower-bound estimate: ½ × 30 > ½ × 0.

upper-bound estimate: ½ × 33 > ½ × 30.

Thus, if the alternative to helping is zero reproductive success, the individual should clearly help. If the alternative is the upper-bound estimate of eggs surviving at the rate of 30%, then it hardly makes any difference whether the bird helps or breeds independently. The real cost is probably between the two, and nearer the lower estimate; the inequality is therefore satisfied, and selection favors helping behavior in young Florida scrub jays. The estimates of both B and C are fallible, however, and the test is uncertain, but it does illustrate how we can attempt a quantitative test of Hamilton's theory.

Florida scrub jays are not the only bird species with helpers at the nest; indeed, the evidence of helping is now expanding rapidly enough to put in doubt the generalization that most bird species breed in monogamous pairs. At present, over 200 bird species with helpers at the nest are known. Some mammal species—such as the red fox, the mongoose, the African hunting dog, and the jackal—have been found to have a breeding unit of one pair with a number of helpers (Figure 10.4). Not all the species with helpers at the nest have the same social system as the Florida scrub jay, however. In some, such as the ostrich (*Struthio camelus*) and another kind of bird called the groove-billed ani (*Crotophaga sulcirostris*), several females lay eggs in the same nest; in the ostrich, only one of the females incubates all the eggs. In other species, such as the dwarf mongoose (*Helogale parvula*), many of the helpers are not genetic relatives of the breeding pair who they help. Helping in mongooses, therefore, cannot be explained by Hamilton's theory.

10.3   Insect societies

10.3.1 The social life of insects

Four groups of insects—ants, bees, wasps and termites—are recognized as social insects (Figure 10.5). The crucial characteristic of these four groups is that, within their nests, they show a reproductive division of labor. Within the nest of a typical ant species there is a single queen who lays nearly all the eggs that are laid in the nest. The rest of the ants are sterile workers. The workers carry out all the duties necessary to keep the colony going, except for the laying of eggs.

There are more than 12,000 social insect species, and they show a fascinating range of ways of life. Within the ants, for example, there are "army" ants with huge colonies of up to 22 million individuals, which bestride the jungle floor eating everything edible in their path. There are "fungus gardening" species that grow fungus on specially prepared rotting leaves and live on the produce of the fungus. Other ant species live by milking honeydew from herds of little insects called aphids. In yet others, workers form living "honeypots," as they hang upside down

(a)

(b)

Figure 10.4

"Helpers" in two mammal species. (a) An adult male grey meerkat (*Suricata suricatta*) in the Kalahari, southern Africa, is babysitting. The offspring are of the dominant female of the group and this babysitter is probably their brother or half-brother. The social behavior of this species has been observed by David Macdonald. The adults not only babysit the offspring of the group—forgoing their own foraging time to do so—but also feed the young and seem to teach them to forage; each youngster attaches itself to one adult for instruction. (b) A pair of female red foxes in England. The females are sisters and living in the same group. In general only one female, the most dominant one of the group, breeds at a time, and some nonbreeding females help to rear the offspring, by giving them food, grooming and playing with them, and retrieving them if they stray from the den. (*Photos: David Macdonald*)

245

Figure 10.5

Polistes, such as this *Polistes gallicus* nesting on an agave, live in relatively simple societies and build small nests.

(Photo: Heather Angel)

from the roof of their nest, their abdomens hugely distended with honey. Australian aborigines dig up the nests, take the ant's head between their fingers, and bite off the honeyed abdomen. The marvels of the social insects are almost endless, but we shall concentrate on some general properties of their social life that are closely related to their altruistic habits.

We must consider ants, bees, and wasps separately from termites. Ants, bees, and wasps belong to the insect order Hymenoptera; termites make up the order Isoptera. A typical hymenopteran colony is founded by a single queen after her "nuptial" flight. If the species is an ant, the foundress bites off her wings and never flies again. She excavates a small nest, lays her first brood, and feeds and rears the larvae. She will never rear larvae again, because this first generation of workers themselves rear the next lot of eggs. It is an important fact that all the workers, in the first and later generations, are females. They are sterile and therefore in a sense sexless, but they contain the genetic makeup of females. Once the colony has reached a certain size the queens lays eggs that are reared as reproductives. The time when reproductives are first produced varies between species. In *Myrmica rubra*, a common garden ant in Europe, it is not until about nine years after founding, when the colony has grown to about 1000 workers.

10.3.2 Altruism in insect societies

The distinctive property of social insects that we need to explain is the sterility of the workers. Why do these workers devote their lives to enhancing the reproduction of another individual? An important part of the answer may be provided by Hamilton's theory. Hymenoptera have an idiosyncratic genetic system, rather different from the usual Mendelian one. They are "haplodiploid." The females, like most animals, have two sets of chromosomes (they are "diploid"); but the males have only one (they are "haploid"). Male offspring inherit no paternal genes; they develop from unfertilized eggs and therefore do not have a father (Figure 10.6). Females contain genes from both their mother and their father.

Haplodiploidy, as Hamilton pointed out, alters the normal pattern of relatedness (Table 10.3). The sisters of one family are exceptionally closely related because they all share exactly the same set of genes from their father; their father has only one set of genes to give. The relatedness between siblings under diploidy was calculated under the premise that if a father contributes a gene to one sibling there is only a chance of one

Figure 10.6
Haplodiploid inheritance. Females have a diploid set of genes but males have only one (haploid) set. Males develop from unfertilized eggs. This kind of inheritance is exceptional, but is found in hymenopteran insects and a few other groups of animals. The relatednesses among individuals differs from those under diploid inheritance: a gene in a male, for instance, has a probability of one of being in his mother (see Table 10.3 and contrast with Table 10.1).

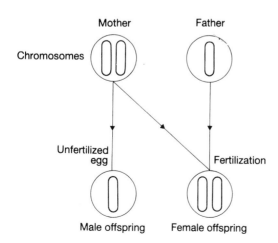

Table 10.3 Relatedness under haplodiploidy. The relatedness is the probability, given a gene is in one kind of individual, that it is in another kind of individual. It can therefore be asymmetrical, as between mother and son.

Relationship		Relatedness (r)
• Mother	daughter	½
• Mother	son	½
• Father	daughter	1
• Father	son	0
• Daughter	mother	½
• Son	mother	1
• Brother	sister	½
• Brother	brother	0
• Sister	sister	¾
• Sister	brother	¼

half that he will give it to another. Under haplodiploidy that chance is one, not a half, and the total relatedness between sisters is therefore (½ × 1) + (½ × ½) = ¾. Because the relatedness of mother to daughter is the same as under diploidy (one half), a female is actually more closely related to her sisters than her daughters, and can, other things being equal, "breed" more efficiently by making sisters than by reproducing daughters. That may be why female hymenopterans have so often evolved sterility. Males, by contrast, have not evolved sterility. But then they are not exceptionally closely related to their siblings. Thus Hamilton's theory explains not only the evolution of sterility in the Hymenoptera, but also its sex bias.

However, the peculiar pattern of relatedness under haplodiploidy is not the only factor leading to the evolution of social behavior. Termites have sterile workers too, but they are not haplodiploid; they are diploid (interestingly, the workers are both male and female in termites—Hamilton's theory does not predict any sex bias to the evolution of sterility in diploid animals). Moreover, there are other groups of animals that have haplodiploid inheritance, but which are not social. Haplodiploidy must be one important factor among several that can lead to the evolution of full social behavior.

10.4   Recognizing relatives

If animals are to act altruistically toward relatives, they must be able to recognize them. How do they do so? No work has been done to answer this question for helpers at the nest such as scrub jays. They might "recognize" as a relative any other birds that grew up in the same nest as themselves. Such birds would usually be siblings, but the actual mechanism, which may be more sophisticated, is unknown. Something is known, however, for other species. In social insects, each colony usually has an individual colony odor. Individuals use the odor to distinguish members of their own colony (who are genetic relatives) from members of other colonies (unrelated) by their characteristic smell. The mechanisms differ according to the species, and we can look at three different systems. In the honeybee (*Apis mellifera*), the specific odor develops at least partly from the diet of the colony; minor differences in the diets of different colonies would give them different characteristic odors. Individual bees learn their own colony odor both from the wax in the honeycomb, and from the continual regurgitation that goes on among the different members of the hive. The hive has a kind of "communal stomach," and individual bees may constantly update their model of the colony odor by this food exchange. Diet may not be the only factor controlling the odor, but its importance has been indicated by experiment. Kalmus and Ribbands moved two hives of honeybees from a typical honeybee environment to an isolated moor that had only one species of flower. The level of fighting between the two hives decreased, probably because they increasingly came to recognize each other as members of the same hive. Likewise, when two parts of one hive were isolated and fed on different diets, the level of fighting within a hive (between the two parts) increased.

Paper wasps (*Polistes*, see Figure 10.5) live in small nests constructed from regurgitated paper. A wasp can distinguish members of her own nest from those from other nests almost immediately after she emerges as an adult. The mechanism is almost certainly the chemical content, particularly the hydrocarbons, in the paper of the nest. The hydrocarbons in the nest are the same as those found in the cuticle (that is, the external surface) of the insects, and individual differences exist between nests because of differences in the hydrocarbons of the wasps who build it. A newly emerged wasp can therefore tell, using her antennae, whether another wasp is from her nest by sensing that wasp's cuticular hydrocarbons. Karl Espelie has obtained several pieces of experimental evidence that implicate the hydrocarbons in the paper structure of the nest. If two nests are treated to remove their hydrocarbons, the members of the two nests lose their ability to discriminate between each other. Likewise, the newly emerging wasps in such a nest can no longer distinguish nestmates from alien wasps (Figure 10.7a). Although in paper wasps a newly emerging individual can already distinguish its nestmates, this is not true of all social insects. In some ants, individuals learn the distinction over time. Young ants of such a species do not challenge other ants in the colony, regardless of their odor; but older ants challenge any ant that does not have the colonial smell. The learning may even take place during a sensitive phase of imprinting.

In several other species, it is thought that individuals may possess a genetic ability to distinguish relatives. The conclusion is suggested when individuals distinguish other members of their species according to their degree of genetic relatedness, even though they have had no prior experience of the individuals they are distinguishing among. The first important experiment was conducted on sweat bees (*Lasioglossum zephyrum*) by Greenberg in 1971. By various genetic tricks, he bred colonies in the laboratory that had twelve different kinds of relatedness to one another, varying from completely unrelated colonies ($r = 0$) to different colonies produced from inbred lines ($r \approx 1$). In his experiments he introduced a bee from one colony to the entrance of another colony. At the colony entrance a "guard" bee generally admits nestmates and challenges foreigners, and Greenberg observed the rate at which guard bees admitted introduced bees bearing differing degrees of relatedness to them (Figure

10.7b). His result was strikingly positive: guard bees admitted bees according to how closely related the introduced bee was, and were more likely to admit more closely related bees. The important part of Greenberg's experimental design was that in nearly all cases the guard bee had never had any opportunity to learn the introduced bee's smell: the guard and the introduced bee had spent all their lives in different colonies. For a minority of cases the bees were separated from the same colony at the larval stage, and some mechanism like that in the paper wasps might have operated, but the positive result still stands even if this minority of cases is ignored. Sweat bees, it appears, can recognize degrees of relatedness to conspecifics of which they have no experience. Since Greenberg's work on sweat bees, a similar ability has been suggested with varying degrees of certainty in other bees and other animals, such as ants, mice, quail, and tadpoles. There are problems in interpreting some of the evidence, but genetic abilities to recognize relatives may prove to be widespread.

10.5   Primate societies

10.5.1 The social life of primates

Different species of primates live in different kinds of societies; indeed, the same species may form different kinds of social groups according to the conditions. The Hanuman langur (*Presbytis entellus*) forms both groups with one male and several females, and multimale groups with several males and a larger number of adult females. The reason is uncertain, but may be related to population density; multimale groups seem to be more common when the total population density of an area is lower. Before considering the altruistic behavior of primates, let us consider briefly some of the different kinds of social groups.

The white-handed gibbon (*Hylobates lar*) lives in the trees of Southeast Asia in monogamous family groups. The male and female are similar in size and appearance, and live as a pair with zero to four young. The pair defend a territory of about half a square mile. Within the family there is no dominance or aggression: male and female live together

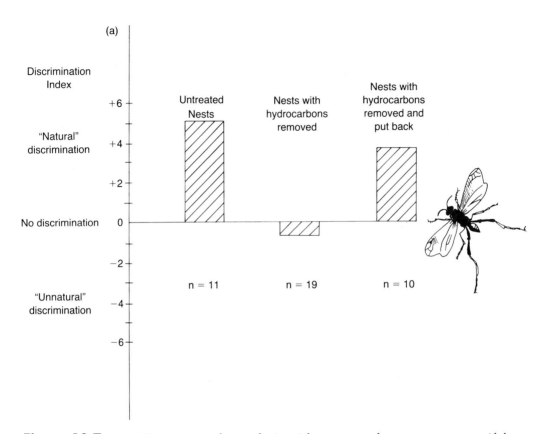

Figure 10.7 (a) Paper wasps do not distinguish nestmates from non-nestmates if the hydrocarbons have been experimentally extracted from the nest structure. Forty nests of *Polistes metricus* were collected in Clarke County, Georgia. The results show whether wasps newly emerged from those nests discriminated between nestmates and non-nestmates. There are three nest treatments: the nest hydrocarbons had been removed, the nest hydrocarbons were first removed and then put back, or the nest was untreated (control). The discrimination index was measured as follows. A newly emerged wasp was put in a chamber with one nestmate and one non-nestmate. The frequencies of "tolerant" and "intolerant" behavior patterns by the newly emerged wasp to the two other wasps was recorded. The discrimination index equals the sum of tolerant behavior to the nestmate plus intolerant behavior to the non-nestmate minus the sum of tolerant behavior to the non-nestmate plus intolerant behavior to the nestmate. The index is positive when the wasp makes the normal (or "natural") discrimination of the species; it is zero when she shows no discrimination between nestmate and non-nestmate; it is negative when she shows reverse (or "unnatural") discrimination, being more tolerant of non-nestmates than nestmates. The histograms show the average behavior of *n* wasps. *(Results of Singer and Espelie)*

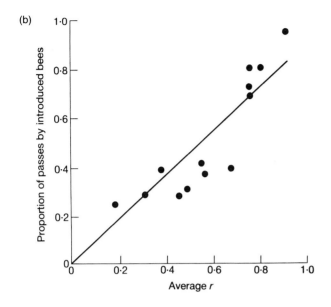

(b)

(b) Whether a guard bee of *Lasiogiossum zephyrum* allows a conspecific bee to enter its nest depends on the genetic relatedness of the two bees: closer relatives are more likely to be admitted. *(After Greenberg)*

peaceably as equals. Other primates, such as the spider monkeys (*Ateles*, Figure 10.8a) also live in family groups. But how different are the societies of hamadryas baboons! The hamadryas baboon (*Papio hamadryas*) is a terrestrial species inhabiting the plains of Northeast Africa and Southwest Arabia. The hamadryas baboon society has several levels of organization. The lowest level, the breeding unit, is a group of 1–10 females with a single male. The male treats the females in the group aggressively, often attacking them if they stray away. The male is about twice as large as the female, which influences the dominance relations of the sexes, but the large size of males is probably an adaptation mainly for fighting off enemies, especially rival males. At the next level, several of these breeding groups may walk around together in larger bands while feeding, and these bands may act as a unit to defend a food source from a rival band. The hamadryas baboon, however, does not defend a territory. The bands do confine their wandering to a large area of about 12–15 square miles, but different bands may overlap in their use of this area. It is therefore called a home range, to distinguish it from a defended area, which would be called a territory. At a higher level, several bands may join into larger groups

Figure 10.8   (a)
Spider monkeys, such
as this family of
Ateles belzebuth
living in Panama, live
in family groups of a
breeding pair and
their offspring. (b)
Howler monkeys
(*Alouatta*), however,
live in groups that
may contain several
adult males and
females. Howler
monkeys, like spider
monkeys, inhabit
Central American
forests. (*Photos: Fritz
Vollrath and Heather
Angel*)

(b)

for sleeping. If suitable shelters for sleeping are difficult to find, as many as 700 hamadryas may sleep together.

Whereas in the hamadryas baboon the mating unit is a group of females aggressively controlled by a single male, in the howler monkey (*Alouatta*), several adult males may live peaceably together in a single group (Figure 10.8b). When Ray Carpenter watched howlers on Barro Colorado Island, in the Panama Canal, the group sizes were variable but contained an average of three males, eight females, and seven young. The male howler monkey is about 30% larger than the female, and has an enlarged voice box covered by a beard. The males roar daily, making the loudest animal noise in the South American forests, a noise that carries for over a mile. This howling serves to space out the different groups. Within each group there is little aggression and no obvious dominance hierarchy.

These three species live in entirely different kinds of societies. One is monogamous, another polygamous within single male groups, another polygamous within multimale groups. One shows dominance and aggression within groups, the other two do not. Two defend territories, the other does not. Other primate species live in yet other kinds of societies. Why are some species aggressive, others not? Some territorial, others not? Our understanding of the diversity of primate societies is still limited. One trend that can be explained is as follows. There is a rough tendency in species in which there are many females per male in the group for the males to be bigger than the females: in monogamous species the male is about the same size as the female and in polygamous species the male is larger. This has presumably arisen because sexual selection has favored larger males in polygamous species, because larger males are more successful in fights over females.

10.5.2 Altruism in primates

Many kinds of altruistic behavior can be seen in primate groups. There is the feeding, carrying, and defense of young, not only by their mothers; cooperative searching for, hunting, and exploitation of food; food sharing; cooperative group defense against enemies; and, most common of all, that favorite pastime of primates, grooming. Most, perhaps all, of these habits should be explained by kin selection. We, however, have considered the application of that theory in two examples already, and will therefore use a primate example to illustrate another reason why altruism may evolve: the theory of reciprocal altruism.

The example concerns "consorting," which is a habit of males in many species that live in multimale groups, whereby a male stays close to a female during the receptive phase of her estrous cycle, and defends her from the advances of other males. It is an adaptation produced by the male competition component of sexual selection. In the olive baboon (*Papio anubis*), Craig Packer observed that two males may occasionally cooperate to fight off a single male: clearly, two will be stronger than one, which makes the advantage of cooperation clear. The advantage, however, is only for the one male that copulates with the female; he has gained a benefit of as much as one extra offspring. The other male has paid a cost of the risk of injury in a fight, but obtains no benefit. He has behaved altruistically. If what we have seen so far were the end of the matter, natural selection should eliminate the altruistic habit. But it is not the end of the matter. The next stage arrives when another female in the troop comes into estrus. The roles of the same two males may now be reversed. Packer saw ten cases in which a male who had previously been "solicited" into cooperating to defend a female (but did not copulate with her) himself solicited a male into cooperative defense. In nine of the ten cases the solicited male was the individual whom he had previously helped. It looks, therefore, as if males form cooperating pairs to defend females and take turns in the copulating that is the end of the defense. If so, it would be an example of reciprocal altruism. Altruism can evolve under individual selection, without any need for the animals to be related, if the altruist is more than paid back later. The danger of any such reciprocal arrangement is that they will be cheated on. There is a clear short-term advantage to receiving altruism but then not paying it back; that way the cheat gains the benefit but does not pay the cost. This being so, reciprocal altruism is expected mainly to evolve in species that form stable groups, with individual recognition. Then a cheat, after gaining a short-term benefit, can be recognized and excluded from future transactions; the cheating will then not pay. Without the opportunity and mechanism of discriminating against cheats, reciprocal altruism is likely to break down. Because reciprocal altruism requires rather special conditions, it may be rarer than kin-selected altruism. It is, however, a theoretical possibility, and in olive baboons it has been realized in fact.

10.6   Manipulated altruism: the control of behavior by parasites

Natural selection, we have seen, normally makes animals behave in their own selfish interests. Even when it favors altruistic behavior, it is only in the interest of some broader form of selfishness. However, conflicts of animals open many opportunities for exploiting other individuals, and animals may not therefore always behave in their own interests. There are probably numerous subtle forms of exploitation within a social group—a possibility we have considered before in relation to animal signals (p. 184)—but the idea of "manipulation" in animal behavior is a relatively recent one, and has not been the subject of much research. For clear examples we are forced to go to those unambiguous situations of conflict, parasite-host relations. Here are many examples in which animals do not behave in their own selfish interests, or the interests of their genetic relatives. Hosts under the influence of parasites do show altruistic behavior, according to the definition used in this chapter, but only because they are in some sense forced to, against their own interest. I should emphasize that I have picked parasite-host examples only because they unambiguously illustrate a process that may be expected to be of wider importance. Many cases of altruistic behavior in nature may be due to even subtler forms of manipulation than those we are about to consider.

A young cuckoo (*Cuculus canorus*) being fed by its foster parent, such as a reed bunting (*Emberiza schoeniclus*), is a striking example of behavioral manipulation by a parasite. It is not in the reed bunting's interest to feed the cuckoo: natural selection favors reed buntings that rear their own offspring, not cuckoos. To our eyes, it is very strange that a reed bunting cannot tell when it is feeding a cuckoo rather than its own offspring. For the insatiable demands of the cuckoo are still met by its tireless foster parents even after the cuckoo has grown larger than its foster parent. It should be easy, we think, for a reed bunting to distinguish a great, ugly cuckoo from a reed bunting. Yet the parent does not. It continues to pour worms down the cuckoo's all-consuming throat. Something about the continual gaping and begging of the cuckoo compels the reed bunting to continue to provide. The reed bunting is being forced by a parasite to do something against its own best interests. The behavior of feeding the young, however, is not abnormal, it is just misdirected.

Let us now move on to some stranger examples, in which the parasite actually changes the behavior of its host. The reasons are to be found in the life cycles of the parasites. Many parasites grow up in a series of different host species. They may start off in one species, be transferred to an intermediate host, and then to a final host, but there can be a problem in effecting the transfer from an intermediate host to the next one. The normal behavior of the intermediate host might not take it near the final host. A kind of fluke (flukes are small, flattened, wormlike animals) called *Dicrocoelium dendriticum,* for example, lives in ants as its intermediate host, and sheep as its final host. How is it to get from an ant to a sheep? The trick is to have the ant eaten by a sheep. Ants normally, understandably enough, avoid being eaten by sheep; they stay down in the soil away from grazing sheep. However, an ant harboring a *Dicrocoelium* changes its behavior. The infected ant typically contains about fifty *Dicrocoelium* individuals. One of these burrows into the ant's brain and somehow causes the ant to climb a blade of grass and fasten its jaws to the top of the blade; it clings fast until eaten, perhaps by a sheep. The parasite has then reached its goal.

There are many fascinating examples of this kind. Here are two more. Nematomorphs (small worm-shaped animals) live in insects as intermediate hosts, but in water as adults. The parasite has to bring its insect to water, and the nematomorphs do somehow bring this about. In one dramatic record, a bee was flying over a pond when suddenly, as it was about six feet above the water, it dived in. As soon as the bee splashed into the water, the worms exploded from the body they had so abused, and swam away, leaving it dead. A final example concerns the parasites of the amphipod *Gammarus lacustris* (Figure 10.9). *Gammarus* is a

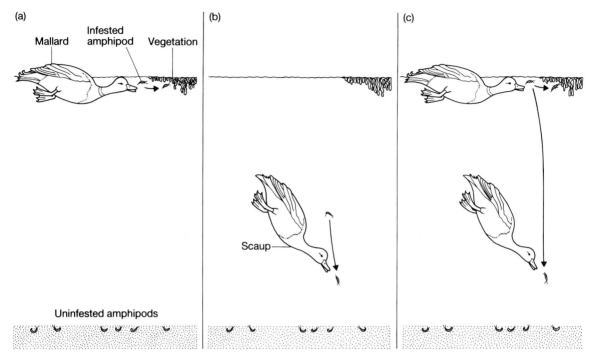

Figure 10.9   Three different parasitic acanthocephalan worms have different effects on the behavior of the amphipod *Gammarus*. Unparasitized *Gammarus* live in the bottom mud. But if parasitized by *Polymorphus paradoxus* (a), they swim toward the light and come to the surface where they may be eaten by dabbling ducks such as mallards (which are in turn parasitized by the *Polymorphus*). If parasitized by *Polymorphus marilis* (b) the *Gammarus* do not swim to the surface, but do come out of the mud; they are fed on by diving ducks such as scaups. (c) *Gammarus* parasitized by *Corynosoma constrictum* swim to the surface but dive when disturbed; they are fed on by both dabbling and diving ducks. *(After Bethel and Holmes)*

small shrimplike animal that lives in freshwater. Normally, *Gammarus* avoid the light, sheltering under pebbles on the bottom. However, when infected by the parasitic worm *Polymorphus paradoxus*, their reaction to light is reversed: they now seek the light and swim just below the surface. The parasite's motive is that the next host after *Gammarus* is a duck that feeds by dabbling at the surface; the *Polymorphus* changes the *Gammarus*'s behavior to make it be eaten by its final host. The duck gets an easy meal, but it infects itself with a parasite when taking it.

Some biologists and psychologists would not include the manipulated altruistic behavior in these parasitological examples as examples of real altruism. However, the dispute is mainly a verbal one, and these examples illustrate how animals are not always in control of their own behavior. Sometimes, unable to help themselves, they behave in the interests of other animals.

10.7　Summary

1. Altruistic behavior poses a paradox for Darwin's theory, which usually predicts that individuals will be selected to leave as many offspring as possible.

2. Group selection is probably too weak a force to explain the existence of altruism.

3. Natural selection may favor altruism under two known conditions: when it is directed toward relatives and when it is paid back later (reciprocal altruism).

4. "Helpers at the nest" in Florida scrub jays is a case of altruism among relatives; measurements in this species enable a quantitative test of the theory.

5. The extreme altruism of sterile workers in social hymenopteran insects may have evolved in part because Hymenoptera have a peculiar genetic system that predisposes females to evolve the sterile worker habit.

6. Social insects can recognize their nestmates (which are relatives) by a learned colony odor, the source of which may be the diet of the colony, or the nest structure itself, depending on the species and conditions. There is evidence for an unlearned, genetic component to kin recognition in sweat bees and other species.

7. Altruistic behavior in primates is often among relatives, but primates also illustrate the theory of reciprocal altruism. Olive baboon males cooperate in pairs, the members of which pay each other back in turn as occasions arise.

8. Altruism can also evolve when it is not in the interests of the altruistic animal. Parasites, such as cuckoos, may manipulate their hosts into behaving in the interest of the parasite but at a cost to the host.

10.8   Further reading

R. Dawkins (1989), Grafen (1991), Krebs and Davies (1983), and Trivers (1985) introduce the subject of altruism; see especially Dawkins, and Williams (1966) for critiques of group selection. R. Dawkins (1981) and Grafen (1982) explain some frequent misunderstandings. On helpers at the nest, see Stacey and Koenig (1990), and Mumme (1992) for the work described in the text. On mobbing of nest predators, see Francis, Hailman, and Woolfenden (1989). Wilson (1971) describes the social insects. On kin recognition, see Singer and Espelie (1992) on paper wasps, and Hepper (1991) and Grafen (1990). On primate social behavior, see Smuts et al. (1986) and Dunbar (1988); for cooperation see Harcourt and De Waal (1992). Packer (1977) is the reference for olive baboons; Wilkinson (1990) describes another classic study of reciprocity, of blood regurgitation in vampire bats. Sherman, Jarvis, and Braude (1992) describe the remarkable social system of the naked mole rat. On manipulation, see R. Dawkins (1982), and Davies and Brooke (1991) on cuckoos.

References

Agosta, W. C. (1992). *Chemical communication: the language of phero-mones*. Scientific American Library, New York.

Andersson, M. (1982). Female choice for extreme tail length in widow bird. *Nature*, 299, 818–819.

Au, W. W. C. (1993). *The sonar of dolphins*. Springer-Verlag, New York.

Axelrod, R. (1984). *The evolution of co-operation*. Basic Books, New York.

Baker, R. R. (1982). *Migration: paths through space and time*. Hodder and Stoughton, London.

Baker, R. R. (1984). *Bird navigation: the solution of a mystery?* Holmes and Meier, New York, and Hodder and Stoughton, London.

Barlow, G. W. (1977). Modal action patterns. In: T. A. Sebeok (ed). *How animals communicate*. Indiana University Press, Bloomington, Indiana, p. 98–134.

Bateson, P. P. G. (ed) (1991). *The development and integration of behaviour*. Cambridge University Press, Cambridge, UK.

Becker, J. B., S. M. Breedlove, and D. Crews (eds) (1992). *Behavioral endocrinology*. MIT Press, Cambridge, Mass.

Bell, R. H. V. (1971). A grazing ecosystem in the Serengeti. *Scientific American*, 226 (1), 86–93.

Belyaev, D. K. (1979). Destabilizing selection as a factor in domestication. *Journal of heredity*, 70, 301–308.

Bentley, D. and R. R. Hoy (1974). The neurobiology of cricket song. *Scientific American*, 231 (2), 34–44.

Berthold, P. (ed) (1991). *Orientation in birds*. Birkhauser, Basel, Switzerland.

Brower, J. V. Z. (1958). Experimental studies of mimicry in some North American butterflies. Part 1. The monarch, *Danaus plexippus,* and viceroy, *Limenitis archippus archippus. Evolution,* 12, 32–47.

Brower, L. P. (1968). Ecological chemistry. *Scientific American,* 220 (2), 22–29.

Cheney, D. L. and R. M. Seyfarth (1990). *How monkeys see the world*. University of Chicago Press, Chicago.

Clutton-Brock, T. H. (1991). *The evolution of parental care*. Princeton University Press, Princeton, New Jersey.

Clutton-Brock, T. H. and S. D. Albon (1979). The roaring of red deer and the evolution of honest advertisement. *Behaviour,* 69, 145–170.

Clutton-Brock, T. H., S. D. Albon, R. M. Gibson and F. E. Guinness (1979). The logical stag: adaptive aspects of fighting in red deer (*Cervus elephas* L.). *Animal Behaviour,* 27, 211–225.

Colgan, P. W. (1989). *Animal motivation*. Chapman and Hall, New York and London.

Colinvaux, P. (1978). *Why big fierce animals are rare*. Princeton University Press, Princeton, New Jersey, and Penguin, London.

Cullen, E. (1958). Adaptations in the kittiwake for cliff-nesting. *Ibis,* 99, 275–302.

Cullen, J. M. (1972. Some principles of animal communication In: R.A. Hinde (ed). *Non-verbal communication*. Cambridge University Press, Cambridge, UK, p. 101–122.

Darwin, C. (1859). *On the origin of species*. John Murray, London. (and many reprints).

Darwin, C. (1871). *The descent of man, and selection in relation to sex*. John Murray, London. (Reprinted 1981 by Princeton University Press, Princeton, New Jersey.)

Darwin C. (1873). *The expression of the emotions in man and animals*. John Murray, London.

Davies, N. B. and M. Brooke (1991). Coevolution of the cuckoo and its hosts. *Scientific American,* 264 (1), 92–98.

Dawkins, M. (1971). Perceptual changes in chicks: another look at the "search image" concept. *Animal Behaviour,* 19, 566–574.

Dawkins, M. S. (1983). The organisation of motor patterns. In: T.R. Halliday and P. J. B. Slater. *Animal behaviour, 1, causes and effects*. Blackwell Scientific Publications, Oxford, UK, p. 75–910.

Dawkins, M. S. (1993). *Through our eyes only?* W. H. Freeman/Spektrum, Oxford, UK.

Dawkins, R. (1981). Twelve misunderstandings of kin selection. *Zeitschrift für Tierpsychologie*, 51, 184–200.

Dawkins, R. (1982). *The extended phenotype.* W. H. Freeman, Oxford, UK. (Paperback: Oxford University Press, New York and Oxford, UK.)

Dawkins, R. (1986). *The blind watchmaker.* W. W. Norton, New York; Longman and Penguin, London.

Dawkins, R. (1989). *The selfish gene,* 2nd edn. Oxford University Press, New York, and Oxford, UK.

Dear, S. P., J. A. Simmons and S. Fritz (1993). A possible neuronal basis for representation of acoustic scenes in auditory cortex of the big brown bat. *Nature,* 364, 619–623.

Dethier, V. G. (1992). *Crickets and katydids, concerts and solos.* Harvard University Press, Cambridge, Mass.

Dunbar, R. I. M. (1988). *Primate social systems.* Cornell University Press, Ithaca, New York, and Croom Helm, London.

Edmunds, M. (1974). *Defence in animals.* Longman, London.

Ewart, J.-P. (1985). Concepts in vertebrate neuroethology. *Animal Behaviour,* 33, 1–29.

Ewert, J.-P. (1987). Neuroethology of releasing mechanisms: prey catching in toads. *Behavioral and Brain Sciences,* 10, 337–403.

Fisher, J. and Hinde, R. A. (1949). The opening of milk bottles by birds. *British Birds,* 42, 347–357.

Francis, A. M., J. P. Hailman and G. E. Woofenden (1989). Mobbing by Florida scrub jays: behavior, sexual asymmetry, role of helpers and ontogeny. *Animal Behaviour,* 38, 795–816.

Frisch, K. von (1967). *The Dance language and orientation of bees.* Harvard University Press, Cambridge, Mass.

Galef, B. G. (1992). The question of animal culture. *Human Nature,* 3, 157–78.

Garcia, J. F., L. P. Brett and K. W. Rusiniak (1989). Limits of Darwinian conditioning. In: B. Klein and R. R. Mowrer (eds). *Contemporary learning theories.* Erlbaum, Hillsdale, New Jersey.

Gould, J. L. (1976). The dance language controversy. *Quarterly Review of Biology,* 51, 211–241.

Gould, J. L. and P. Marler (1987). Learning by instinct. *Scientific American,* 256 (1), 74–85.

Grafen, A. (1982). How not to measure inclusive fitness. *Nature,* 298, 425–426.

Grafen, A. (1990). Do animals really recognize kin? *Animal Behaviour,* 39, 5–31.

Grafen, A. (1991). Modelling in behavioural ecology. In: J. R. Krebs and N. B. Davies (eds). *Behavioural ecology: an evolutionary approach,* 3rd edn. Blackwell Scientific, Boston, Mass., and Oxford, UK, p. 5–31.

Greene, E. (1987). Individuals in an osprey colony discriminate between high and low quality information. *Nature,* 329, 239–241.

Griffin, D. (1992). *Animal minds.* University of Chicago Press, Chicago.

Guilford, T. (1993). Homing mechanisms in sight. *Nature,* 363, 112–113.

Hailman, J. P. (1969). How an instinct is learned. *Scientific American,* 221 (6), 98–106.

Harcourt, A. H. (1992). Coalitions and alliances: are primates more complex than non-primates? In: A. H. Harcourt and F. B. M. De Waal (eds). *Coalitions and alliances in humans and other animals,* Oxford University Press, New York, p. 445–471.

Harcourt, A. H. and F. B. M. De Waal (eds) (1992). *Coalitions and alliances in humans and other animals.* Oxford University Press, New York, and Oxford, UK.

Harden Jones, F. R. (1968). *Fish migration.* Edward Arnold, London.

Hepper, P. G. (1991). *Kin recognition.* Cambridge University Press, New York, and Cambridge, UK.

Hollis, K. L. (1990). The role of Pavlovian conditioning in territorial aggression and reproduction. In: D. Dewsbury (ed). *Contemporary issues in comparative psychology.* Sinauer, Sunderland, Mass, p. 197–219.

Hollis, K. L., E. L. Cadieux and M. M. Colbert (1989). The biological function of Pavlovian conditioning: a mechanism for mating success in the blue gourami (*Trichogaster trichopterus*). *Journal of Comparative Psychology,* 103, 115–121.

Huber, F. and J. Thorson (1985). Cricket auditory communication. *Scientific American,* 253 (6), 46–54.

Ioalè, P., M. Nozzolini, and F. Papi (1990). Homing pigeons do extract directional information from olfactory stimuli. *Behavioral Ecology and Sociobiology,* 26, 301–305.

Keverne, E. B. (1992). Primate social relationships: their determinants and consequences. *Advances in the Study of Behaviour,* 21, 1–37.

King, B. J. (1991). Social information transfer in monkeys, apes, and hominids. *Yearbook of Physical Anthropology,* 34, 97–115.

Konner, M. (1982). *The tangled wing,* Holt, Rinehart, & Winston, New York. (Reprinted Penguin, London, 1993.)

Krebs, J. R. and N. B. Davies (1993). *An introduction to behavioural ecology,* 3rd edn. Blackwell Scientific Publications, Boston, Mass., and Oxford, UK.

Krebs, J. R. and G. Horn (eds) (1991). *Behavioural and neural aspects of learning and memory.* Oxford University Press, Oxford, UK.

Lehrman, D. S. (1964). The reproductive behavior of ring doves. *Scientific American,* 211 (6), 48–54.

Lohmann, H. (1992). How sea turtles navigate. *Scientific American,* 266 (1), 100–106.

Lorenz, K. (1958). The evolution of behavior. *Scientific American,* 199 (6), 67–78.

Lorenz, K. (1965). *Evolution and modification of behavior.* University of Chicago Press, Chicago.

Lorenz, K. (1966). *On aggression.* Harcourt, Brace, and World, New York, and Methuen, London.

Manning, A. and M. S. Dawkins (1992). *An introduction to animal behaviour,* 4th edn. Cambridge University Press, Cambridge, UK.

Marler P. (1959). Developments in the study of animal communication. In: P. R. Bell (ed). *Darwin's biological work: some aspects reconsidered.* Cambridge University Press, Cambridge, UK, p. 150–206.

Maynard Smith, J. (1991). Honest signalling: the Philip Sidney game. *Animal Behaviour,* 42, 1034–1035.

McCleery, R. H. (1983). Interactions between activities. In: T. R. Halliday and P. J. B. Slater (eds). *Animal behaviour 1, causes and effects.* Blackwell Scientific Publications, Oxford, UK, p. 134–167.

McLannahan, H. M. C. (1974). Some aspects of the ontogeny of cliff nesting behaviour in the kittiwake (*Rissa tridactyla*) and the herring gull (*Larus argentatus*). *Behaviour,* 44, 36–88.

Michelsen, A., B. B. Andersen, W. H. Kirchner and M. Lindauer (1989). Honeybees can be recruited by a mechanical model of a dancing bee. *Naturwissenschaften,* 76, 277–280.

Møller, A. P. (1988). Female choice selects for male sexual tail ornaments in the monogamous swallow. *Nature,* 332, 640–642.

Møller, A. P. (1989). Viability costs of male tale ornaments in a swallow. *Nature,* 339, 132–134.

Møller, A. P. (1990). Effects of a haematophagous mite on the barn swallow (*Hirundo rustica*): a test of the Hamilton and Zuk hypothesis. *Evolution,* 44, 771–784.

Morton, E. S. and J. Page (1992). *Animal talk: science and the voices of nature.* Random House, New York.

Mumme, R. L. (1992). Do helpers increase reproductive success: an experimental analysis in the Florida scrub jay. *Behavioral ecology and sociobiology,* 31, 319–328.

Nottebohm, F. (1989). From bird song to neurogenesis. *Scientific American,* 260 (2), 74–79.

Packer, C. (1977). Reciprocal altruism in *Papio anubis. Nature,* 265, 441–443.

Papi, F. (ed) (1992). *Animal homing.* Chapman & Hall, London.

Ray, J. C. and S. M. Sapolsky (1992). Styles of male social behavior and their endocrine correlates among high-ranking wild baboons. *American Journal of Primatology,* 28, 231–250.

Reeve, H. K. and P. W. Sherman (1993). Adaptation and the goals of evolutionary research. *Quarterly Review of Biology,* 68, 1–32.

Ridley, M. (1993). *Evolution.* Blackwell Scientific Publications, Boston, Mass., and Oxford, UK.

Ridley, M. (1995 [in press]). The expressionist theory of the emotions. In: B. Shore and C. Worthman (eds). *The emotions.* Cambridge University Press, New York.

Roeder, K. D. (1965). Moths and ultrasound. *Scientific American.* 212 (4), 94–102.

Roper, T. J. (1983). Learning as a biological phenomenon. In: T. R. Halliday and P. J. B. Slater (eds). *Animal behaviour. 3. Genes, development, and learning.* Blackwell Scientific Publications, Oxford, UK, p. 178–212.

Roper, T. J. and S. Redston (1987). Conspicuousness of distasteful prey affects the strength and durability of one-trial avoidance learning. *Animal Behaviour,* 35, 739–747.

Rusak, B., H. A. Robertson, W. Wisden, and S. P. Hunt (1990). Light-pulses that shift rhythms induce gene expression in the suprachiasmatic nucleus. *Science,* 248, 1237–40.

Sapolsky, R. M. (1990). Stress in the wild. *Scientific American* 262 (1), 116–123.

Sapolsky, R. M. (1993). *Why zebra don't get ulcers.* W. H. Freeman, New York, and Oxford, UK.

Schaller, G. B. (1972). *The Serengeti lion,* University of Chicago Press, Chicago.

Schneider, D. (1974). The sex-attractant receptor of moths. *Scientific American,* 231 (1), 28–35.

Schnitzler, H.-U. and J. Ostwald (1983). Adaptations for the detection of fluttering insects by echolocation in horseshoe bats. In: J.-P. Ewert, R. R. Capranica and D. J. Ingle (eds), *Advances in vertebrate neuroethology.* Plenum Press, New York, p. 801–27.

Seyfarth, R. M. and D. L. Cheney (1992). Meaning and mind in monkeys. *Scientific American,* 267 (6), 122–128.

Shear, W. A. (ed) (1986). *Spiders: webs, behavior, and evolution.* Stanford University Press, Stanford, California.

Sherman, P. W., J. U. M. Jarvis and S. H. Braude (1992). Naked mole rats. *Scientific American,* 267 (3), 72–78.

Sherry, D. F. and Galef, B. G. (1984). Cultural transmission without imitation milk bottle opening by birds. *Animal Behaviour,* 32, 937–938.

Shettleworth, S. J. (1983). Memory in food-hoarding birds. *Scientific American* 248 (3), 102–110.

Singer, T. L. and K. E. Espelie (1992). Social wasps use nest paper hydrocarbons for nestmate recognition. *Animal Behaviour,* 44, 63–68.

Smith, H. G. and R. Montgomerie (1991). Sexual selection and the tail ornaments of North American barn swallows. *Behavioral Ecology and Sociobiology,* 28, 195–201.

Smith, J. N. M. (1974 a,b). The food searching behaviour of two European thrushes, 1,11. *Behaviour,* 48, 276–302, and 49, 1–61.

Smuts, B. B., D. L. Cheney, R. M. Seyfarth, R. W. Wrangham and T. T. Struhsaker (eds) (1986). *Primate societies.* University of Chicago Press, Chicago.

Sparks, J. (1982). *The discovery of animal behaviour.* Little Brown, Boston, and Collins/BBC, London.

Stabell, O. B. (1984). Homing and olfaction in salmonids. *Biological Reviews,* 69, 333–338.

Stacey, P. B. and W. D. Koenig (1990). *Cooperative breeding in birds.* Cambridge University Press, New York.

Staddon, J. E. R. (1983). *Adaptive behavior and learning.* Cambridge University Press, Cambridge, UK.

Stephens, D. W. and J. R. Krebs (1986). *Foraging theory.* Princeton University Press, Princeton, New Jersey.

Stoddart, D. M. (1990). *The scented ape.* Cambridge University Press, Cambridge, UK, and New York.

Suga, N. (1990). Biosonar and neural computation in bats. *Scientific American,* 262 (6), 60–68.

Timberlake, W. (1993). Animal behavior: a continuing synthesis. *Annual Review of Psychology,* 44, 675–708.

Tinbergen, N. (1951). *The study of instinct.* Oxford University Press, Oxford, UK, and New York.

Tinbergen, N. (1953). *Social behavior in animals.* Methuen, London.

Tinbergen, N. (1958). *Curious naturalists.* Country Life, London. (2nd edn., 1974, Penguin Books, London.)

Tinbergen, N. (1963). On aims and methods of ethology. *Zeitschrift Für Tierpsychologie,* 20, 410–433.

Tomasello, M., A. C. Kruger and H. H. Ratner (1993). Cultural learning. *Behavioral and Brain Sciences*, 16, 495–552.

Trivers, R. L. (1985). *Social evolution*. Benjamin/Cummings, Menlo Park, California.

Turkkan, J. S. (1989). Classical conditioning: the new hegemony. *Behavioral and Brain Sciences*, 12, 121–179 [with commentary continuing in subsequent volumes].

Turner, J. R. G. (1977). Butterfly mimicry: the genetical evolution of an adaptation. *Evolutionary Biology*, 10, 163–206.

Vines, G. (1981). Wolves in dogs' clothing. *New Scientist*, 91, 640–652.

Waterman, T. H. (1989). *Animal migration*. Scientific American Library, New York.

Wenner, A. M. and P. M. Wells (1990). *Anatomy of a controversy: the question of a "language" among bees*. Columbia University Press, New York.

Whiten, A. (ed) (1991). *Natural theories of mind*. Blackwell, Oxford, UK.

Wingfield, J. C., R. E. Hegner, A. M. Duffy and G. F. Ball (1990). The "challenge hypothesis": theoretical implications for patterns of testosterone secretion, mating systems, and breeding strategies. *American Naturalist*, 136, 829–46.

Index